SUNLIGHT TO ELECTRICITY

Prospects for Solar Energy Conversion by Photovoltaics

SUNLIGHT TO ELECTRICITY

Prospects for Solar Energy Conversion by Photovoltaics

Joseph A. Merrigan

The MIT Press

Cambridge, Massachusetts, and London, England

PUBLISHER'S NOTE

This format is intended to reduce the cost of publishing certain works in book form and to shorten the gap between editorial preparation and final publication. The time and expense of detailed editing and composition in print have been avoided by photographing the text of this book directly from the author's typescript.

The MIT Press

Printed in the United States of America

Second printing, March 1976
Third printing, July 1977

Library of Congress Cataloging in Publication Data

Merrigan, Joseph A
Sunlight to electricity.

Bibliography: p.
Includes index.
1. Solar batteries. 2. Photovoltaic power generation. I. Title.
TK2960.M47 338.4'7'62131213 75-6933
ISBN 0-262-13116-1

CONTENTS

ILLUSTRATIONS

PREFACE

This book is an assessment of the economic and technological prospects for commercial development of photovoltaic material into a viable converter of sunlight into electrical energy. The first chapter is devoted to defining U.S. energy demand and supply until the year 2000, and illustrates the long-term incentive to utilize solar energy. The extent of solar energy falling on the United States in relationship to energy demands is treated in chapter two. Chapters three and four deal with the technology and state-of-the-art in direct conversion of sunlight to electricity. The remaining chapters treat the economics of photo electricity generation, market considerations and projections, and technological and business forecasts to the end of this century.

I am particularly grateful to Professors Henry D. Jacoby, Edward B. Roberts and James W. Meyer at the Massachusetts Institute of Technology Sloan School of Management, and Energy Laboratory, for their advice in researching this subject and presenting the results. The integrity of this work was enhanced by the kind and considerate sharing of insights and research by: K. Boer, University of Delaware; J. Berkowitz, A. D. Little, Inc.; E. Berman, Solar Power Corp.; L. Ephreth, Boston College; P. Fang, Boston College; R. Grutsch, Jr., Solar Division International Harvester Co.; P. Glaser, A. D. Little, Inc.; D. Jewett, Tyco Laboratories, Inc.; R. LaRose,

A. D. Little, Inc.; W. Morrow, Jr., M.I.T. Lincoln Laboratory; and D. Trevoy, Eastman Kodak Company. I very much appreciate the time and effort spent by these executives and scientists in discussing the technological and business aspects of producing photovoltaic energy conversion systems for various applications.

Chapter 1

EXPECTED ENERGY USAGE IN THE UNITED STATES TO THE YEAR 2000

Introduction

One of the most important factors in the economic development of the United States is the availability of an adequate supply of power to maintain and improve its mechanized society. The availability of conveniently usable energy is a result of the many available fossil fuel resources, the ability to convert these resources into easily distributed and usable forms, and a willingness to invest in natural resources such as nuclear fission processes. In 1972 the U.S. supplied about 87% of its energy needs from domestic sources.[1] However, for a combination of reasons, not the least of which is a continuing growth in demand, this percentage is expected to decrease in the immediate future. Presently, with 6% of the world population, the U.S. uses one third of the world energy production.[2] Over the decade 1960-1970, the U.S. use of energy rose from 4.5×10^{16} to 6.8×10^{16} Btu/yr., an increase of 4.3% compounded annually. A

---

1. Battelle Research Outlook, "Our Energy Supply and Its Future," Editor; A. B. Westerman, Battelle, Columbus, Volume 4, Number 1, 1972, p. 3.

2. Joint Committee on Atomic Energy, "Certain Background Information for Consideration When Evaluating the National Energy Dilemma," U.S. Printing Office, Washington, 1973.

continuing yearly increase in demand for energy through 1985 between a high of 4.5 and a low of 3.3% per year is expected.[3] If the supply is to increase concomitantly problems and challenges must be faced such as; offshore oil development, power plant siting, nuclear power safety, coal mine safety, sulfur and particulate emission control, strip mining, increased prices and increased quantities of imported oil and gas, development of new energy sources, and a host of other environmental, political sociological, and economic factors.

This book is concerned with the potentiality that the solar energy will be used to significantly enhance the supply of conveniently usable energy within this century. More specifically, it deals with the direct production of electricity through absorption of sunlight by appropriate semiconductors in which this photo energy is effectively converted into electrical energy—the photovoltaic effect. This method circumvents the lengthy processes of conversion of sunlight by photosynthesis to stored energy in plants, fossilization, recovery, burning, and using the heat to power electricity generators. It is also a more direct method than using solar energy to heat liquids

---

3. National Petroleum Council, "U.S. Energy Outlook, A Summary Report of the National Petroleum Council," December 1972, p. 15.

or gases for running generators. The basic objectives of this study are:

1. To illustrate the scope of energy requirements in the U.S. until 2000 and the relative magnitude of solar energy available.
2. To review and clarify the scientific principles of photovoltaic energy conversion as related to its potential for technological development.
3. To determine the state-of-the-art and problems which limit massive utilization of photovoltaic converters.
4. To realistically estimate the trends in economic and technological parameters which will influence the development of photovoltaic conversion.
5. To project the probable range of business opportunities in the manufacture of photovoltaic converters in the next 25 years.

Energy Demand

Detailed studies of the expected demand for energy in the U.S. in the future have been made recently.[4,5,6,7]

---

4. National Petroleum Council, "U.S. Energy Outlook, A Report of the National Petroleum Council's Committee on U.S. Energy Outlook," December 1972.

5. Shell Oil Company, "The National Energy Outlook," March 1973.

6. U.S. Department of the Interior, "United States Energy Through the Year 2000," December 1972.

The predictions of energy demand up to 1980 seem reasonable barring major long term changes in national or international policies. Energy patterns are already committed in many ways in that most of the electrical capacity that can be functioning by 1980 has already been ordered; every major mass transit system that can be functioning has been ordered; and the consumption of energy by automobiles and for space heating can be predicted with a high degree of confidence.[2] The main factors which might affect them would be a sustained shortage of importable petroleum, as in the winter of 1973-1974, which could cause an awareness of the societal dependence upon energy and the need to conserve it. This and the concomitant increase in price of energy could decrease the demand, although the latent demand may still be present if price and supply conditions similar to the past decade were re-established. The predictions of demand from 1980 to 2000 are rather uncertain because of the many factors which can change over the long term. It is significant that by 1980 nearly 50% of the U.S. oil requirements may have to be imported, i.e., 22% of the total energy demand.[2,8,9]

---

7. Associated Universities, Inc., Report "Reference Energy Systems and Resource Data for Use in the Assessment of Energy Technologies," April 1972. Report to U.S. Office of Science and Technology, under Contract OST-30; Document AET-8.

Projections of the energy demand to the year 2000 are shown in Figure 1. This figure is a modification of fold-out M of reference 2 which summarizes the projections in references 4,5,6,7,8. Projection A represents an average growth of 4.2% per year. Projection B, C, and D correspond to the projections of reasonable high, intermediate, and low growth rates estimated by the Shell Oil Company,[5] and F is by the Department of the Interior.[6] The projections by the National Petroleum Council actually were made only through 1985 and the extensions of 2000 represent about a 3.3% annual growth rate. Projection B was based upon an annual growth rate of 4.5% 1970-1981, 4.3% 1981-1985, and about 3.3% 1985-2000. In projection D there was assumed an annual growth rate of 3.5% 1970-1981, 3.3% 1981-1985, and about 2.9% 1985-2000. The annual demand volume is expressed in British thermal units (Btu) which is the quantity of heat required to raise the temperature of one pound of water one degree Fahrenheit. This unit of energy is used because of its universality. The other units such as tons of coal, barrels of oil, etc., can easily

---

8. National Petroleum Council, "U.S. Energy Outlook, An Initial Appraisal 1971-1985," July 1971.

9. Lawrence Livermore Laboratory, "Energy: Uses, Sources, and Issues," UCRL-51221 May 30, 1972.

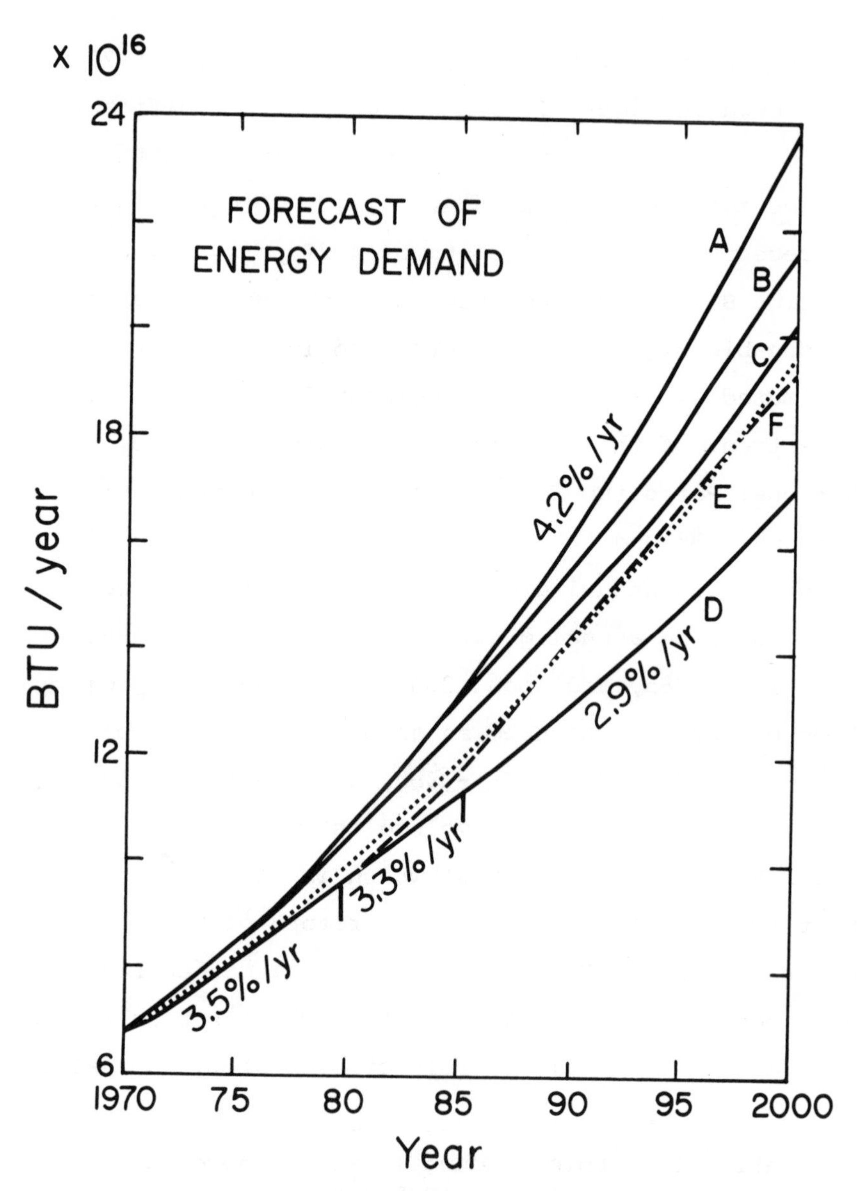

Figure 1. U.S. Energy Demand Forecast. (Data from Ref. 2)

be converted into their Btu equivalents as follows:

Crude Oil:	1 Barrel (Bbl)=42 gallons (gl)= 5,800,000 Btu
Electricity:	1 KiloWatt hour (kWhr)= 3,412 Btu
Coal:	1 Ton = 26,000,000 Btu (average)
Natural Gas:	1 Cubic Foot = 1,032 Btu

In arriving at the majority of the above projections, the most significant long range determinants of energy demand were deemed to be: (1) economic activity (GNP); (2) cost of energy; (3) population; and (4) environmental controls. Of course, these are not the only determinants, but these were found to explain most of the past changes in energy demand.[4] Factors which may become very important, such as supply limitations and political decisions, are not reflected in the projections shown in Figure 1. A substantial reason for the lowest estimate of demand (curve D) is the assumption that consumers will improve the efficiency in which energy is used by better home insulation, increased engine fuel economy, better machinery,...etc. The impetus for this would be the increased costs of energy and the normal technological developments.

The basic assumptions for the intermediate growth rate, curve C, from 1970-1985 were:

(a) Average annual growth in real GNP of 4.2%.

(b) Average annual growth in population of 1.1%.

(c) Energy used for environmental development increasing from 2% in 1970 to about 5% of our total usage in 1985.

(d) Negligible change in real prices for energy.

(e) Improved technology for fuel substitution.

(f) No capital limitations or other restrictions on total energy supplies.

The above parameters are not independent variables and it should be obvious that if (f) were not true and energy shortages arose; (a), (c), and (d) could be affected. During 1974 both (a) and (b) were lower, and (d) higher, than the above assumptions. Hence, considerable judgment must go into deductions based upon these projections. Sensitivity analyses for energy demand with regard to the major assumptions are available in reference 4.

The demand for energy may be subdivided into major use areas to get better insight into future trends. The studies referred to so far are not all in strict agreement even on historical data, but do not disagree significantly for purposes of this discussion. Figure 2 is derived from fold out B of reference 2 and illustrates the flows of energy into electrical energy generation and the residential and commercial, industrial, transportation, and nonenergy sectors during 1970. The units are arbitrary and the numbers represent the percentage of total input primary energy (before exports and field use) employed in that particular phase of energy utilization. About 22% of the energy went into residential and commercial uses predominantly for conditioning the indoor atmosphere. Twenty-nine percent went to industry, 23% to transportation, 6% to nonenergy (synthesis of petrochemicals, fertilizers, plastics,...etc.), and 21% to electricity generation. It may be noted from Figure 2 that the transportation and electrical energy generation sectors are very wasteful. About 75% of fuel which goes into transportation is wasted through inefficient processes. Sixty-five percent of the primary energy which goes to produce electricity, a secondary but very convenient form of energy, is lost. By the time the electricity is utilized and partially lost through inefficiencies, this figure is even higher. The expected average annual growth rates[4] in these sectors from 1970 to 1980 under assumptions for energy demand C in Figure 1 are as follows:

Residential and Commercial; 3.4%

Industrial; 2.9%

Transportation; 3.9%

Electricity Generation; 6.9%

Nonenergy; 5.1%

This indicates that the largest growth rate will be in a very inefficient sector; electricity generation. This is largely due to the relative ease of distribution of electricity to the residential/commercial, and industrial sectors, the high demand for the convenience

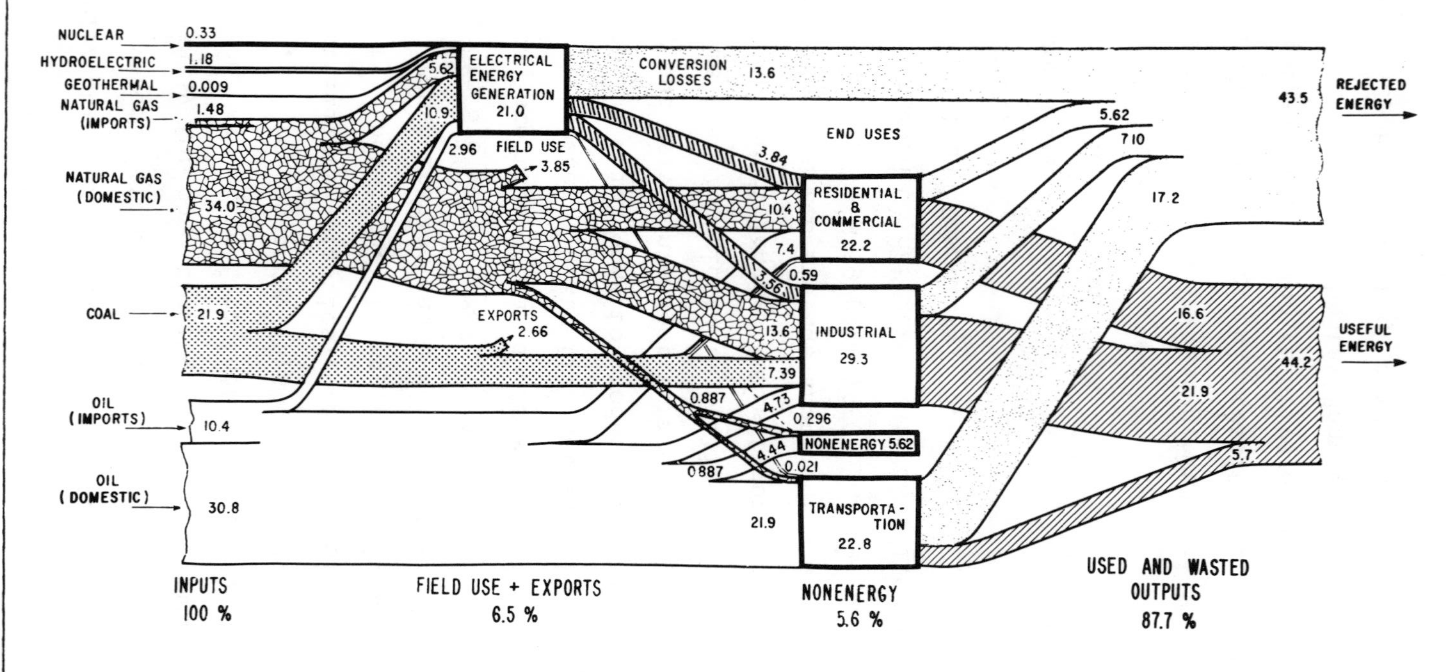

Figure 2. Normalized U.S. Energy Flows in 1970. (Data from Ref. 2)

of electricity, and its usage in pollution abatement. The substitutability of primary fuels in electricity production also makes it probable that electricity may be an increasingly utilized form of "end use" energy so users do not have to change their facilities from, say oil burning to coal burning as the relative costs and supplies change. The expected energy use "mix" from 1960 to 1985 is illustrated in Figure 3 which was derived from illustrations in reference 2. This implies a relatively high expected dependence upon electricity in the near future. The percentage of total energy used within the U.S. for power purposes which is devoted to electricity generation will increase from about 25% in 1970 to nearly 36% in 1985.[10] Although direct end use of primary fuels will increase considerably, electricity generation will account for an increasing proportion of the total primary fuel usage. This projection is consistent with past experiences. From 1950 to 1970 electricity consumption increased[11] at an average annual growth rate of 7.5%. In 1970 about 55% of the 5.1 x 10^{15} Btu of electricity generated was used in the residential

---

10. National Petroleum Council, "Guide to National Petroleum Council Report on the United States Energy Outlook," December 1972.

11. S. H. Schurr, "Energy Research Needs," Resources for the Future, Inc., Washington, D. C., October 1971, p. I-6.

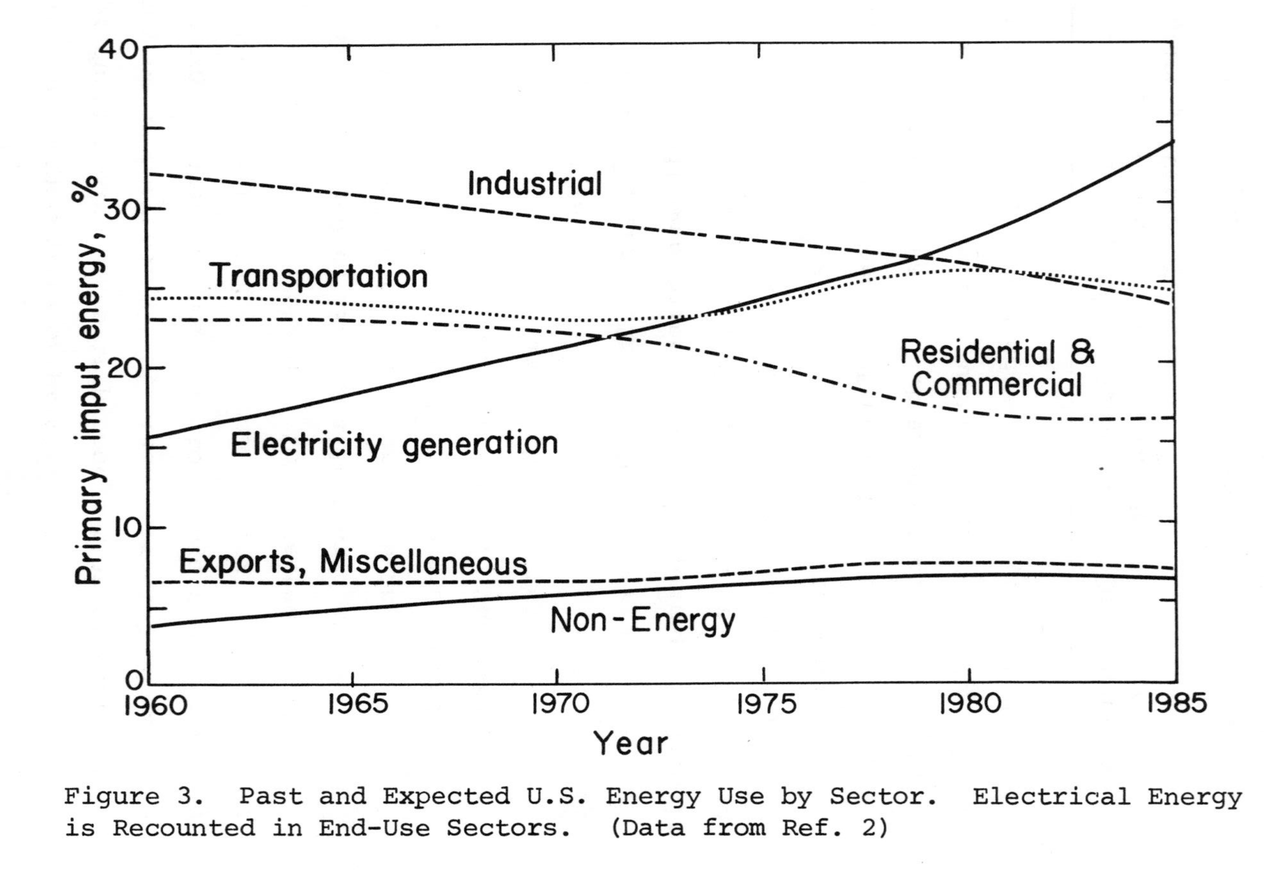

Figure 3. Past and Expected U.S. Energy Use by Sector. Electrical Energy is Recounted in End-Use Sectors. (Data from Ref. 2)

and commercial sector while the rest went into industry.[4] This percentage is expected to remain nearly constant through 1985, with a relatively small amount going into transportation by 1985. It would appear then that the product of photovoltaic energy conversion, electricity, will be in increasing demand in this century.

Energy Supply

In 1972 the United States' energy consumption[12] of 7.5×10^{16} Btu was supplied from the following primary fuels; nuclear 1%, hydro 4.1%, coal 18.6%, natural gas 31.8%, and oil 44.4%. About 84-87% came from domestic sources. Thirty-two percent (1.8×10^{9} Bbl) of the oil and 4.4% (10^{12} ft^{3}) of the natural gas was imported. This represents an increasing dependence upon oil stimulated in large measure by pollution abatement and the Clean Air Act.[13] The increasing dependence upon imports and recent large increases in prices makes predictions of supplies rather tenuous. However, many studies[4,5,6,7,8,9,10,11] have been devoted to predictions of what the domestic supplies can be under various price and regulation policies. As a result of historical developments and policies, and

12. British Petroleum Company, "1972 Statistical Review of the World Oil Industry."

13. D. C. White, *Technology Review*, *76*, No. 2, 11 (1973).

the long delay periods involved in producing primary fuels from new discoveries and producing electricity from nuclear sources, it is probable that domestic supplies will not be able to meet demand in an economically efficient fashion for the rest of this century. Hence, there probably will be considerable importation of energy with concomitant outward financial flows, or shortages.

Estimates of the U.S. resources of coal, petroleum, natural gas, uranium, geothermal energy and oil from oil shale are made periodically by the U.S. Geological Survey.[14] Accuracy of the estimates range from 20 to 50% for identified recoverable resources to about an order of magnitude for undiscovered-submarginal resources. Estimates are affected considerably by assumptions of market price and technological capability. As price increases, estimates of recoverable resources increase because it becomes economically feasible to extract low grade or hard-to-recover sources of energy from the earth. Less expensive and more efficient recovery processes also tend to increase the quantities considered recoverable. The coal resources are estimated at 3.2×10^{12} tons (8.3×10^{19} Btu), about 1000 times our present annual

14. P. Theobald, S. Schweinfurth, and D. Duncan, "Energy Resources of the United States," Geological Survey Circular 650, Washington 1972.

energy usage. Of this, 2—3.9 x 10^{11} tons are considered identified and recoverable presently. Production of coal is expected to increase by as much as 6.7% annually from 0.59 x 10^{9} tons in 1970 to about 1.6 x 10^{9} tons in 1985.[4] About one third of the latter figure would be used for exportation and making synthetic gas and liquids. It would appear that the coal supplies could last at least 100 years and possibly reduce the reliance on imported energy.

The total resource base[14] for petroleum liquids is estimated at 2.9 x 10^{12} barrels (1.7 x 10^{19} Btu). However, only 5.2 x 10^{10} Bbl is identified and recoverable at the price and technological conditions of 1972. The National Petroleum Council's estimates are somewhat lower at 0.81 x 10^{12} Bbl resource base. Annual production is expected to go from 4.1 x 10^{9} Bbl in 1970 to 5.7 x 10^{9} Bbl in 1985 under reasonably favorable development conditions. If present conditions prevail, this production may be as little as about 3.8 x 10^{9} Bbl in 1985. In this case, only 20% of the energy demand would be met by domestic oil while about 38% would have to be met by importation.[10]

The 1972 U.S. Geological Survey estimates that the U.S. resources in natural gas total 6.6 x 10^{15} ft^3 (6.8 x 10^{18} Btu) of which 0.29 x 10^{15} ft^3 is identified and recoverable. If the conditions under which 2.2 x 10^{13} ft^3 were produced in 1970 prevail, the production in 1985 may fall to about 1.5 x 10^{13} ft^3.

However, under favorable exploration, price and regulation conditions,[4] the production could be as high as 3.1×10^{13} ft^3.

The demand for nuclear fuels is derived from demand for electricity. As more electricity is demanded and if environmental and safety restrictions are as at present with respect to coal and nuclear fuels, a sharp increase in the demand and respondent supply of uranium oxide will be in store for the rest of the century. In conventional deposits where U_3O_8 is the major product there are about 1.6×10^6 tons of which 2.5×10^5 are identified and recoverable.[14] In 1970 about 7×10^9 Watts of electricity was supplied by nuclear generation. This is expected[15,16] to grow to about 3×10^{11} by 1985. The National Petroleum Council projects that this will grow to 9.8×10^{11} by 2000. By this projection the annual usage of U_3O_8 in 1985 would be 7.1×10^4 tons and the total reserves used up to that time would be about 5×10^5 tons. The heat input provided by the 7.1×10^4 tons in 1985 would be about 2×10^{16} Btu (a relatively small portion of demand) of which two-thirds would be lost in generating 2×10^{12} kWhr of electricity. Thus, it can

---

15. Atomic Energy Commission, "Nuclear Power Growth 1971-1985," Washington, December 1971.

16. Federal Power Commission, "The 1970 National Power Survey Part 1," December 1971.

be seen that more exploration for minerals and research in nuclear processes for electricity generation will be required if the hoped-for growth and sustaining of a major nuclear power industry is to be achieved. Most experts believe that this will develop and the breeder reactor will make nuclear power generation much more efficient with respect to mineral usage.

The potential geothermal energy sources are thought to exceed 4×10^{19} Btu of which only 10^{16} can be considered identified and recoverable.[14] Considerable exploration, deep drilling, and research and development will be required for geothermal energy to contribute significantly to the U.S. needs by 2000.

Oil shale could possibly produce about 1.5×10^{20} Btu. It is not economically feasible at present to extract the oil, but if prices increase and stay at the 8-10 dollar per Bbl region, up to 6×10^{11} Bbl (3.5×10^{18} Btu) could be considered identified and recoverable.[14] Under the most favorable conditions, about 2.7×10^{8} Bbl (1.5×10^{15} Btu) may be produced[4] in 1985. However, water used in the process of oil extraction may limit production over the long term. Tar sands from which oil may be extracted are in very small quantities in the U.S. and are not expected to contribute significantly to the domestic energy supply. Canada has vast tar sands resources which are beginning to be developed.

Hydroelectric energy presently accounts for about 4% of the U.S. needs. Because most of the suitable dam sites have been developed, the growth of hydroelectric power will be only 1.6% per year[4] until 1985 as smaller sites are utilized. The portion of the U.S. energy needs it will supply will decrease from 4% in 1970 to 3% in 1985. It may be used in conjunction with other electricity generating means, however, as a means of energy storage (pumped water storage). In off-peak hours the fossil or nuclear powered generators will provide electricity for pumping water upward, and in peak load hours the water will be used to run additional turbines. This and combined-cycle (Brayton-Rankine) plants which utilize the presently wasted hot exhaust from gas turbines to generate steam for conventional steam-electric generators will serve to increase the efficiency of electricity generation and storage. The combined-cycle plants which are expected by 1985 will use about 30% less fuel[4] than conventional plants built in 1972.

Other unconventional and undeveloped sources of energy may be utilized by 1990 to 2000 but are unlikely to be significant contributors to the U.S. energy supplies before 1985. These include use of nuclear fusion, fuel cells, magnetohydrodynamics, thermionic devices, and agricultural energy. Solar energy, as will be discussed in more detail, is a bountiful source which may compete economically with more "conventional" sources within this century.[17]

The supply of energy to meet the U.S. demand is illustrated in Table 1. The trends in supplies of energy to meet U.S. demand from 1970 to 2000 are depicted in Figure 4 for one set of assumed conditions[18] which seem reasonable and in near agreement with other sources.[4,5] It would appear that much dependence is expected on nuclear processes and importation. This Atomic Energy Forum[18] estimate was developed as a maximum which would be feasible if many present limitations are diminished and a massive national effort is made to develop nuclear systems. The term imports and/or shortages is used to describe the expected deficit between demand and domestic supply because there is some concern that the national economy may not support outward monetary flows for energy of over about $20 billion per year. At a "real" price of $8 per Bbl of foreign crude oil, this would be 2.5 billion Bbl per year; 1.5×10^{16} Btu, about double the 1970 level.

---

17. W. E. Morrow, Jr., Technology Review, 76, No. 2, 31 (1973).

18. Joint Committee on Atomic Energy, Atomic Industrial Forum Informal Report on March 7, 1973, (see reference 2).

Table 1. Estimated Energy Resources* in the United States and 1970 Usage

Resource	Estimated Resources*, Btu	1970 Usage, Btu
Petroleum	1.7×10^{19}	2.1×10^{16}
Natural Gas	6.8×10^{18}	2.2×10^{16}
Coal	8.3×10^{19}	1.3×10^{16}
Hydropower	4×10^{15} per yr.	0.3×10^{16}
Nuclear	5×10^{17} (plus)	0.02×10^{16}
Geothermal	4×10^{19}	0.0007×10^{16}
Shale	1.5×10^{20}	negligible
Tar Sands	minor	negligible
Solar	5×10^{19} per yr.	negligible
Total Domestic 1970		5.9×10^{16}
Imported Oil		0.8×10^{16}
Imported Gas		0.1×10^{16}
Total used 1970		6.8×10^{16}

*Estimates from references 4 and 14, 1970 usage from references 2 and 4. The total estimated resources would only be developable under considerably higher price conditions and then not completely recoverable. Economic conditions and capacity related to energy production in 1970 allowed only the production listed: some had to be imported.

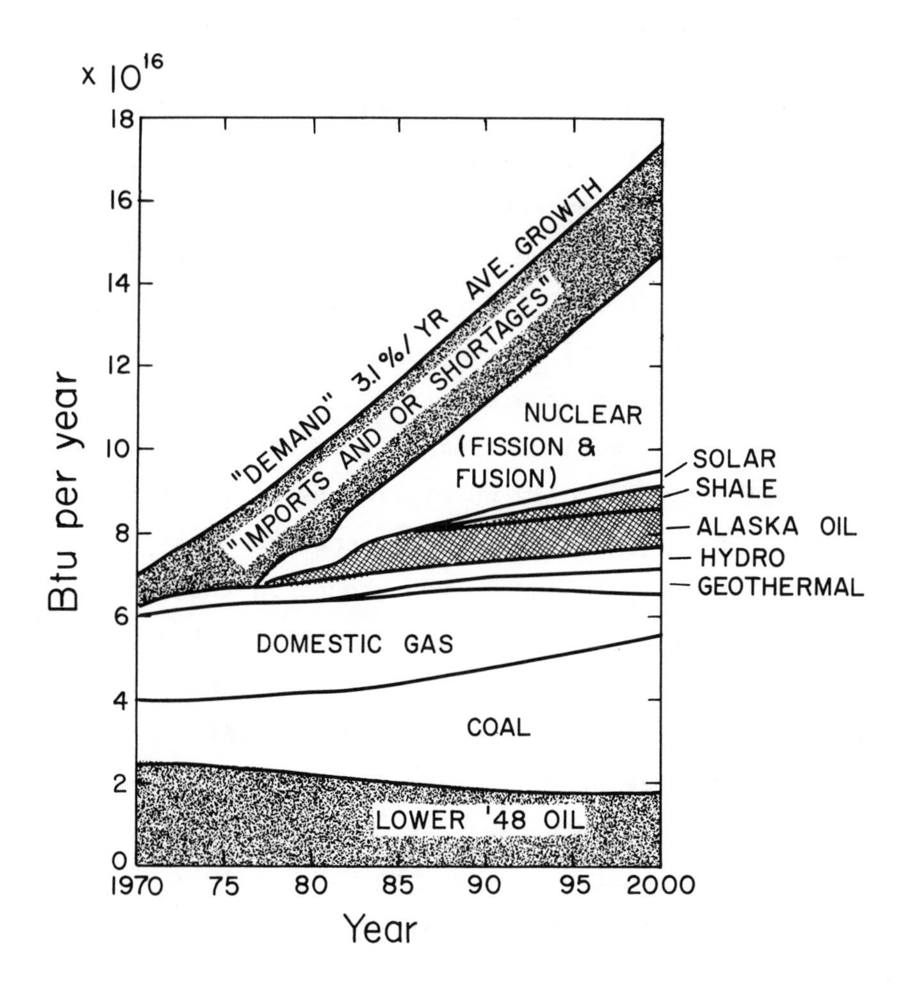

Figure 4. Expected Trends in U.S. Energy Supplies Until 2000. (Data from Refs. 2 and 18)

Chapter 2

SOLAR ENERGY AS A RESOURCE

It is the sun which has made possible fossil fuels via the long natural processes of storage of solar energy in plants and subsequent decay processes. The sun has been and will be a long term energy source for the earth, lasting for billions of years as we know it. The best estimate of the energy received from the sun by the earth's atmosphere is 3.75×10^{21} Btu per year.[1] Present world consumption of energy is 2.2×10^{17} Btu per year, only 60 millionths of the incoming energy. The yearly average incidence of solar energy in near-earth space and on the ground in the continental U.S. is 130 and 17 Watts per ft^2, respectively.[2] The contiguous 48 states comprise about 3.1×10^6 square miles. This multiplied by $(5280\ ft/mi)^2$, 365.25 d/yr., 24 hr/d, 17 W/ft^2, and 3.412 Btu/Whr, yields 4.4×10^{19} Btu per year. This is greater than 600 times the energy used by the U.S. in 1972 (discounting natural uses of solar energy).

Many scientific proposals have been made for harnessing this energy.[3] The methods differ greatly

---

1. See reference 13, Chapter 1.

2. NSF/NASA Solar Energy Panel, "An Assessment of Solar Energy as a National Energy Resource," University of Maryland, December 1972.

3. See reference 17, Chapter 1.

in details but basically involve conversion of sunlight into thermal energy for space heating or generation of steam to make electricity, and direct conversion of sunlight to electricity via the photovoltaic effect. The preponderance of proposals are for ground-level capture of sunlight. This necessitates energy storage schemes because of the variations in sunlight intensity from day to night, and weather and seasonal changes. Figure 5 (derived from reference 4) illustrates the distribution of yearly average and December average solar energy which falls on a horizontal surface in various regions of the 48 states. The southern and especially the southwestern portions of the U.S. receive the most sunlight and hold most promise for cost-effective solar energy utilization. Figure 6 illustrates the average monthly sunshine hours in selected cities in these regions.[5] The incident solar energy in the U.S. is least during December, January, and February. If the average solar energy incidence during this period were 10 W/ft^2, a typical residential roof area of 1000 ft^2 would be exposed to an average of 10 kW of energy; the equivalent of about 90 Amps in a 110 volt circuit or about 13 horsepower (746 W/Hp). It would nearly double in

---

4. M.I.T. Energy Laboratory and M.I.T. Lincoln Laboratory, "Proposal for Solar-Powered Total Energy Systems for Army Bases," Massachusetts Institute of Technology, July 1973.

5. H. Landsberg, H. Lippmann, Kh. Paffen, and C. Troll, "World Maps of Climatology," Springer-Verlag, New York, 1965, Edition 2.

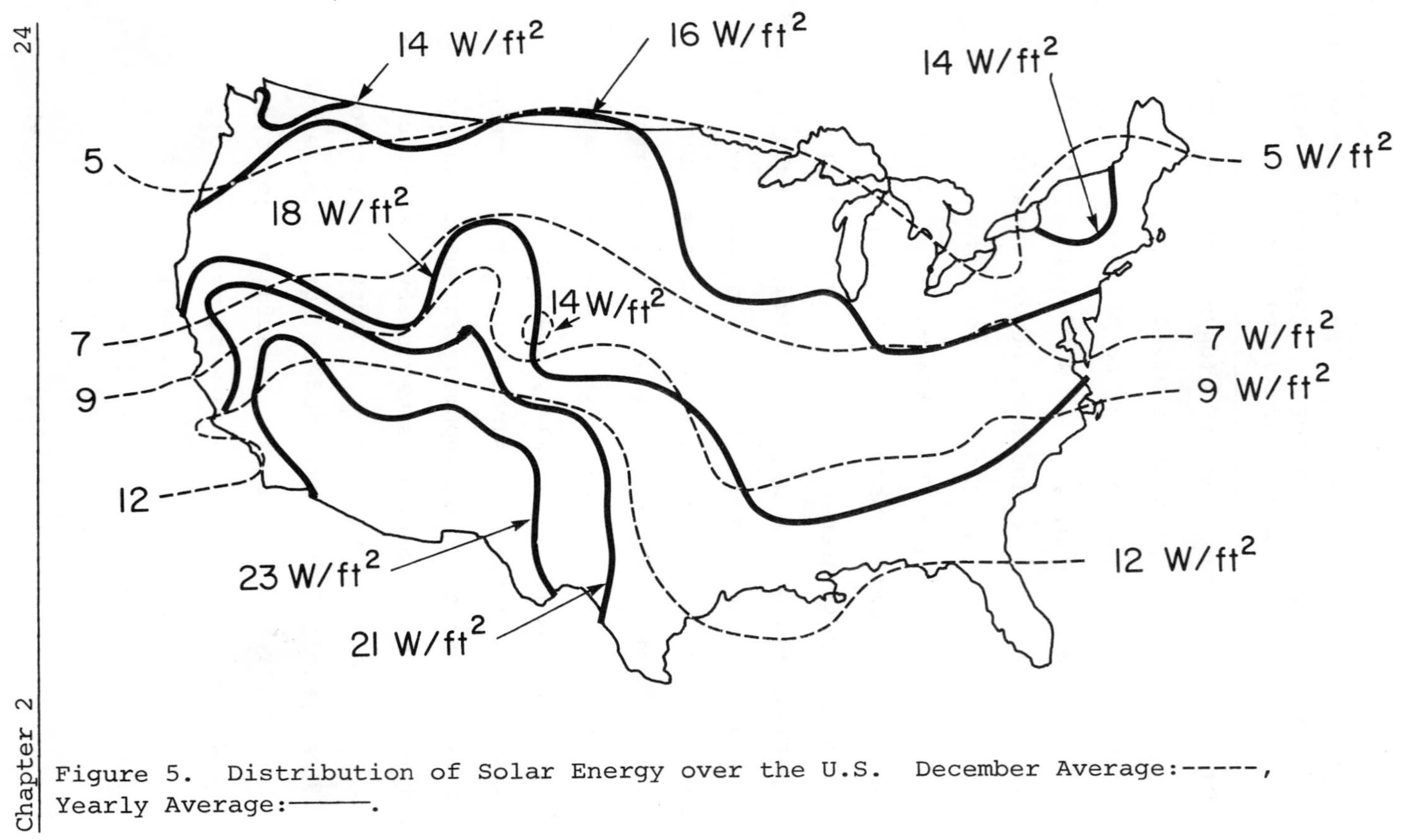

Figure 5. Distribution of Solar Energy over the U.S. December Average:-----, Yearly Average:———.

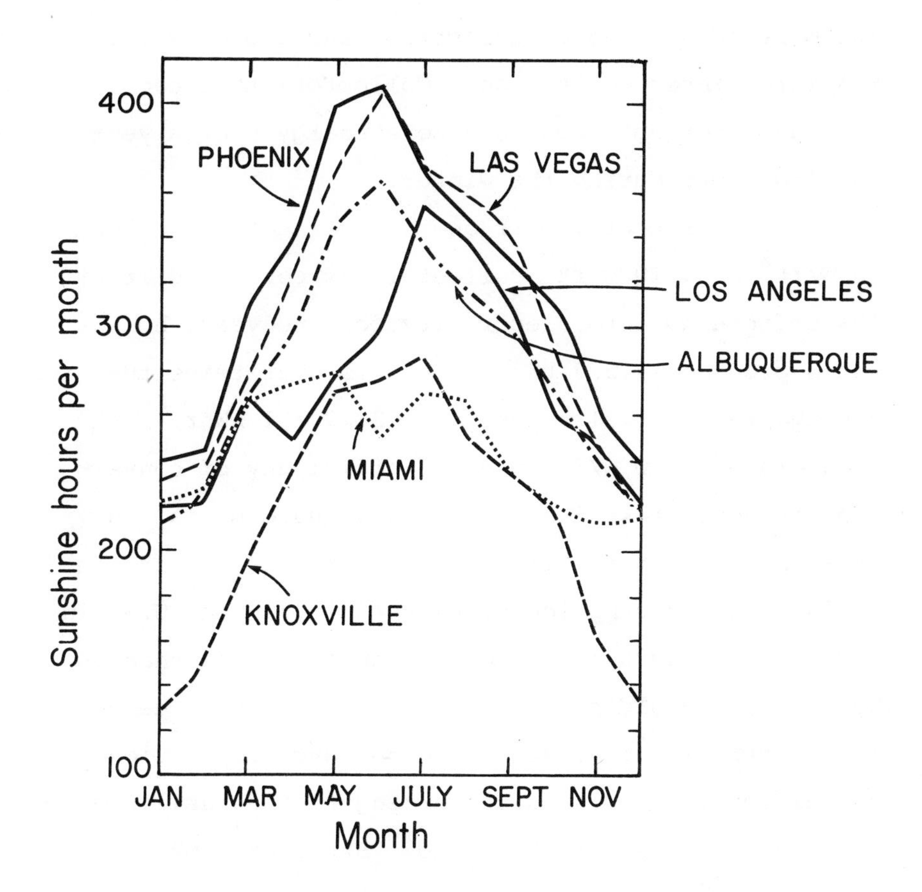

Figure 6. Average Monthly Sunshine in Selected U.S. Cities. (Data from Ref. 5)

the summer. Tipping a previously horizontal collector to face the south at 45° above horizontal yields about 30% more annual energy collection and approximately 80% more winter collection. Collectors that can track the sun intercept about 80% more energy over a year and 150% more during the winter.[4]

To add perspective, if on the average one collected 20 W/ft^2 on a 1000 ft^2 collector and converted it at 10% efficiency into electricity for one year, he would produce 17,532 kWhr. This is over twice the average residential usage in 1972; 7,691 kWhr.[6] At an average of 20 W/ft^2 and 10% efficiency of conversion it would take less than 1350 square miles (an area about 37 by 37 miles) to provide for all of the residential consumption of electricity in the U.S. in 1972. About 4090 square miles, an area less than 64 by 64 miles would provide for <u>all</u> of the U.S. demand for electricity in 1972. With the same collection and efficiency factors, and converting to the basic Btu heat unit, the entire 7.2×10^{16} Btu energy usage of the U.S. could be provided by sunlight falling on 43,000 square miles, an area about 208 by 208 miles, less than 0.012 of the total U.S. area (3,615,000 $mile^2$).

---

6. Edison Electric Institute, "Statistical Year Book of the Electric Utility Industry for 1972," New York, November 1973.

Neglecting natural uses of sunlight, the most commonly used method of converting sunlight to conveniently usable power is by absorbing it in black materials which are thus heated. This thermal energy is transferred to water or gas flowing through the black absorber. Millions of small solar water heaters are used in Israel, Australia, and Japan for residential hot water supplies. A well designed heater will have one or two glass or plastic windows over the absorber to reduce convection and infrared radiation losses. Of the 80—85% of the light which is transmitted through the windows, 80—95% is absorbed by the black absorber with a resultant 65—80% collection of the incident energy. Losses inside the collector from convection, conduction, and re-radiation bring net collection efficiencies down to 50-75%. With such efficiencies, water and space heating of homes can be practical at least in parts of the country.[7] More sophisticated systems involving parabolic collectors and higher temperatures may compete with fossil fuel burning for providing heat and electricity to industrial parks[3] and in base load electricity for public utilities. Generally a 60% solar collection efficiency and a 40% efficiency for steam powered turbines are assumed, with a resultant 24% overall efficiency for electricity production. All of these systems involve

---

7. K. W. Boer, Chem. Eng. News, January 29, 12 (1973).

some provision for storage of heat for periods of 3-4 days to allow continuous power release for utilization.

Direct photovoltaic conversion of sunlight to electricity is a very enticing prospect for clean production of electrical power. The main limitations presently are low efficiency of conversion and high costs of solar cells. A ground-based large scale plant may be envisioned to convert solar energy to electricity with about a 12% efficiency. However, some form of energy storage is necessary for continuous production. Electricity could be used to electrolyze water and generate hydrogen (about 95% efficiency) which could be stored and piped to consumers for direct use or for converting back to electricity via a fuel cell (about 80% efficient). The overall efficiency would be about 9%. In cost and efficiency, ground-based photovoltaic power stations presently do not compare favorably with solar powered thermal power stations.

An alternative scheme[8] for providing large scale electricity supplies from solar cells is to build the solar array in synchronous orbit where there is a nearly continuous supply of sunlight of about 8 times

---

8. P. E. Glaser, "Space Resources to Benefit the Earth," Third Conference on Planetology and Space Mission Planning, The New York Academy of Sciences, October 1970.

the average intensity on the ground. Using microwaves the transmission efficiency from direct current in space to direct current on earth[9] would be 55-75%. Then overall efficiency might be in the 5-10% region before distribution of the electricity.

There are many technical and cost considerations which relate to the utilization of photovoltaic energy conversion as a "pollution-free" means of meeting the energy demands of the U.S. Technical progress is being made, energy demand is increasing, and the costs of primary fuels are increasing at an unprecidented rate. The rest of this study will deal with the present technological state-of-the-art, cost considerations, and the markets for photovoltaic arrays. It will be shown that even though sunlight as a resource presently is not being harnessed as effectively as it could be, technological progress will probably make it a well utilized direct source of electrical energy in the future.

---

9. W. C. Brown, "Microwave Power Transmission in the Satellite Solar Power Station System," Raytheon Company Technical Report ER72-4038, 1972.

Chapter 3

PRINCIPLES OF PHOTOVOLTAIC ENERGY CONVERSION

Materials and Solid State Mechanisms

Most solar cells are composed of crystalline semiconductors prepared to have solid state characteristics which promote separation of photocarriers and a resultant flow of electricity in an attached external circuit. Although some are made from cadmium sulfide (CdS), cadmium telluride (CdTe), gallium arsenide (GaAs), germanium (Ge), and other combinations, the predominant basic material is silicon (Si). This is the substrate employed in most of the solar arrays used to power earth satellites. It is well defined theoretically and experimentally.

Each atom in a pure silicon crystal has four valence electrons which are shared with adjacent silicon atoms in covalent bonding. If the crystal is doped with an impurity such as phosphorus which occupies the same lattice sites but has five valence electrons, the doped crystal contains valence electrons in excess of a pure crystal—one for each of the phosphorus atoms. At a given temperature many of these excess electrons are separated from the phosphorus atoms by thermal energy and are free to wander in the crystal making it an electron-conducting, n-type, semiconductor. If silicon is doped analogously with boron which has only three valence electrons, there will be one electron too few to complete the covalent bonding in the vicinity of each boron atom. This electron vacancy, or absence of an electron, appears to the lattice to be positively charged because an electron would normally occupy that site. The electron vacancy, or hole, is a positive charge carrier which when thermally detached from the boron impurity is free to wander in the crystal making it a hole-conducting, p-type, semiconductor. Because positive nuclear charges balance the negative valence electron charges in both p and n-type semiconductors, there is no macroscopic charge disequilibrium in or on the crystals.

If the p-type and n-type crystals were figuratively joined together to make a perfect single crystal, the electrons in the n-type portion would diffuse across the joining boundary (p-n junction) into the "electron deficient" p-type region and the holes would likewise diffuse into the "hole deficient" n-type region until a voltage equal to the sum of the diffusion potentials of the holes and electrons was established across the p-n junction. Thus a permanent electric field is established in the region of the junction. In practice junctions like this are made by ion implanting, diffusing, or otherwise growing p-type impurities into an n-type crystal or n-type impurities into a p-type crystal.

The resultant energy band structure of interest is shown schematically in Figure 7. By virtue of the electron and hole flows described above, the conduction and valence bands in the p-type material have risen relative to those in the n-type region. Because the Fermi level which represents the electrochemical potential[1] was originally higher in the n-type crystal the level is equalized within the "joined crystal" and across the p-n junction by virtue of the n-type material feeding electrons to the p-type material and vice versa with respect to holes. The p-type portion now contains a disproportionate amount of electrons and is at a more negative potential (higher in the diagram) which causes a field across the junction. This causes electrons which are photolytically generated in this region to flow toward the n-type portion and photolytically generated holes to flow toward the p-type region. If light of greater energy than the valence to conduction band gap falls upon the crystal, it may be absorbed by valence band electrons which are thus excited to the conduction band leaving a vacant electronic state, a hole, in the valence band. Under the influence of the electric field the photo-excited electrons will be driven toward a lower energy state in the n-type portion of the crystal while the holes move toward a lower energy state which for them is in the p-type region. The photocarriers have moved into respective regions of the crystal where like charges

---

1. J. Tauc, "Photo and Thermoelectric Effects in Semiconductors," Pergamon Press, New York, 1972, p.18.

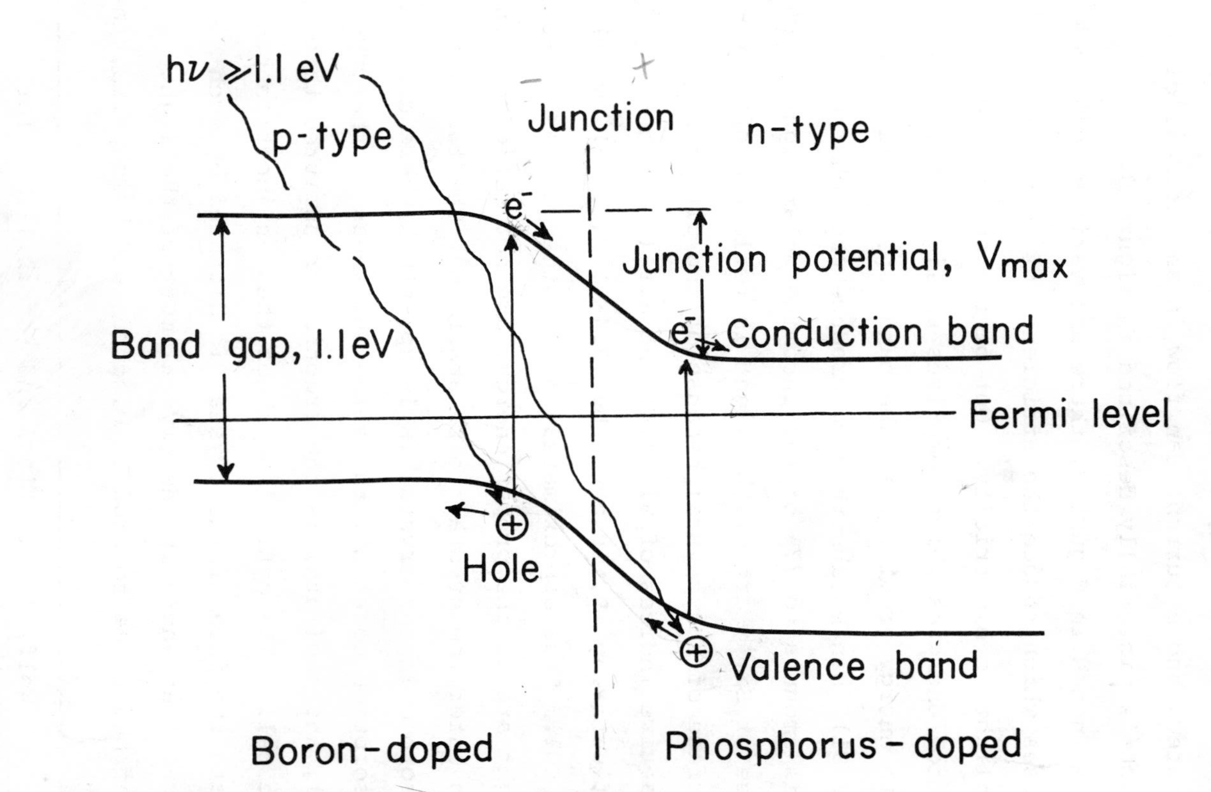

Figure 7. Energy Band Structure at a p-n Junction in Silicon.

are the majority charge conductors, a photovoltage is created, and a current can flow in an external circuit which is pictorially described in Figure 8.

As noted in Figure 8, there are various loss processes which reduce the efficiency of conversion of sunlight to electricity. Out of the spectrum of photon energies which comprises sunlight only those with energy greater than the band gap, E_g, (1.1 eV for Si)[2] can be absorbed to produce photocharges. This amounts to 77% of the incoming solar light in the case of Si. Part of the light is reflected off the surface of the crystal because of the differing refractive indices of Si (3.4) and air (1). The reflectivity, R, is given by $(3.4-1)^2/(3.4+1)^2$ which is 30%. Thus approximately 50% of the photons in sunlight are of the proper energy, are not reflected, and can enter the crystal to be absorbed. The fraction absorbed in the crystal is $1-\exp(-\alpha th)$, there α is the absorption constant, th is the thickness of the crystal, and $\exp(-\alpha th)$ is the fraction transmitted through the cell. The optimal depth of the p-n junction beneath the face of the crystal is dictated by such factors as where in the crystal most of the light is absorbed so as to produce electron and hole pairs, the

2. M. Wolf, "Limitations and Possibilities for Improvements of Photovoltaic Solar Energy Converters," Proc. IRE, 48, 1246 (1960).

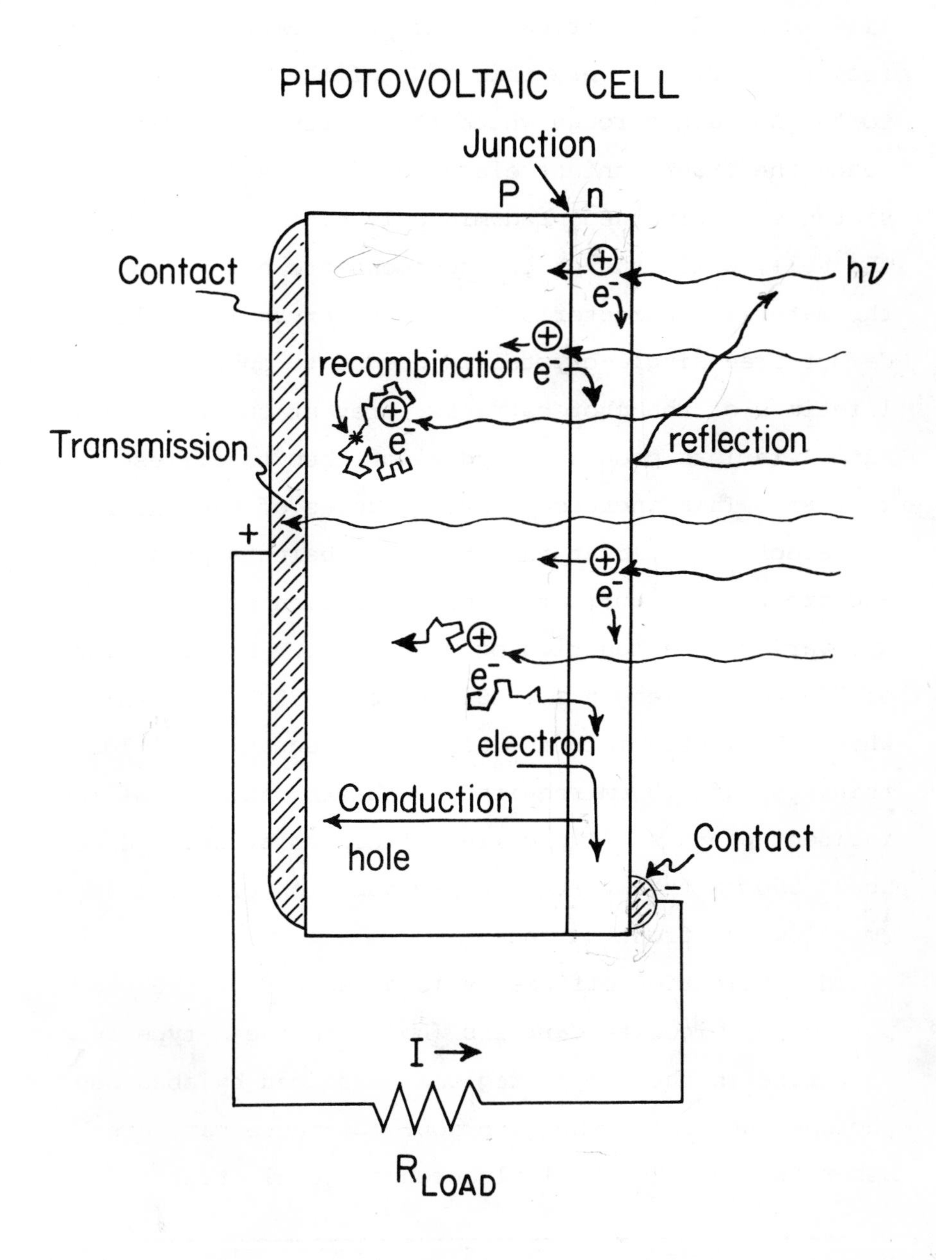

Figure 8. Schematic Representation of Light Interactions and Current Flows in a Photovoltaic Cell.

lifetimes and mobilities of the photocarriers, and the resistance of the very thin side of the junction next to the surface through which the current must flow to reach the front contact electrode.[3] This latter resistance is governed predominantly by the geometry of the cell, while the other phenomena are a function of the material characteristics of the crystal itself. A defect free single-crystal is needed to enhance the lifetimes of the photocarriers, prevent their recombination (a loss process), and allow them to diffuse further. This increases their chances of encountering the electric field at the junction, being separated, and producing a current. The absorption coefficient, α, varies as a function of wavelength. For red light of 700 nm wavelength α in Si is about 2000 cm^{-1} which when substituted into the absorption equation yields a thickness of 3.5 μm required to absorb one half of the incident photons. Typically silicon solar cells are about 250 μm thick with the p-n junction 0.5 to a few μm below the front surface.

The collection efficiency is a measure of the proportion of minority carriers (holes in the n-type and electrons in the p-type regions) produced by absorbed photons that reach the junction. Of those carriers generated outside the influence of the electric

---

3. J. F. Elliott, "Photovoltaic Energy Conversion," in Direct Energy Conversion, Edited by G. W. Sutton, McGraw-Hill Co., New York, 1966, p1-37.

potential at the junction, some diffuse toward the junction while others diffuse away and recombine in the bulk of the cell or at the surface. In typical silicon cells the collection efficiency ranges from 60-80%.

Another loss process involves conducting electricity through the very thin resistance layer between the p-n junction and the front side electrical contact. The geometric factors of contact location, say using a grid or screen instead of a spot contact, and junction depth which partly governs the resistance, have to be balanced against loss of incident light through masking and collection efficiency deterioration. Typically the efficiency of the cell is reduced several percent by this resistance.[4]

The voltage which can be developed by a solar cell is a function of the excess of minority carriers on each side of the p-n junction. This voltage is less than the band gap because of junction losses. The voltage developed increases with intensity of illumination toward a condition where the minority carrier density approaches the majority carrier density and the voltage approaches the band gap energy. The voltage could never go beyond that because the junction

---

4. P. Rappaport and J. Wysocki, "The Photovoltaic Effect," in Photoelectronic Materials and Devices, Edited by S. Larach, Van Nostrand Co., New York, 1965, pp. 239-275.

potential would be nullified. In reality it never approaches the band gap energy very closely because as soon as the junction potential (Figure 7) is counterbalanced by separated photocarriers, no internal field exists and the open-circuit voltage is reached. This voltage is directly proportional to the band gap energy. The junction loss (about 0.5 V in Si; 45%) decreases exponentially with increasing band gap. It also decreases as the temperature decreases because of a decrease in thermally generated carriers which may reach the junction. The maximum open circuit voltage of an experimental silicon cell is about 0.6 V, while under full sunlight at ambient temperature and maximum power conditions the voltage is about 0.4 V. Thus, even on a single photon absorbed basis the cell converts a relatively small portion of the original energy ($>$ 1.1 eV, the band gap) to useful energy. Theoretical limits on conversion efficiency involve other factors also and will be considered later in this chapter.

Spectral Considerations

Sunlight - The basis for photovoltaic energy conversion is the absorption of photons of appropriate energy in a properly selected semiconductor. In selecting the semiconductor one must consider the spectrum of the light source relative to the absorption and reflection spectra of the material. The fraction of energy in the light source which is subject to absorption should be maximized within the geometrical and solid state physics and chemical constraints on the process. The light source under consideration is the sun.

The extraterrestrial spectrum of sunlight is close to that of a black body radiator at 5800°K.[5,6] However, because the atmosphere absorbs a disproportionate amount of short wavelength light, it is generally accepted that the shape of the solar spectrum at sea level most closely fits that of a 6000°K radiator in the wavelength region below 1250 nm.[7] The coarse sunlight spectra above the atmosphere[8] and at sea level[9] are shown in Figure 9. The insert in Figure 9

---

5. R. M. Goody, "Atomispheric Radiation," Clarendon Press, Oxford, 1964, pp. 417-426.

6. C. W. Allen, Quart. J. Roy. Met. Soc., 84, 307 (1958).

7. S. T. Henderson, "Daylight and Its Spectrum," American Elsevier Publishing Co., New York, 1970.

8. M. Nicolet, Arch. Met. Geophys. Biokl., B3, 209 (1951).

9. D. Diermendjian and Z. Sekera, Tellus, 6, 382 (1954).

illustrates the percent transmittance of the atmosphere[10] as a function of wavelength from 400-700 nm.

The predominant portion of the energy in sunlight lies in the visible region. At sea level with the sun overhead in a clear sky at normal humidity, the total radiant flux is reduced to about 80% of that above the atmosphere.[7] Presently accepted values of the energy distribution in the spectrum are:

	$\leq$400 nm	400-800 nm	800-1000 nm	$\geq$1000 nm
Extraterrestrial:	9%	49%	13%	29%
Sea Level:	7%	58%	14%	21%

Under these conditions the energy falling on the atmosphere is 130 W/ft^2 and at sea level; 104 W/ft^2. Considering silicon as an absorbing photovoltaic material with a band gap of 1.1 eV one can calculate the energy which can be absorbed. If a nonreflective but transmitting coating over the surface reduces the reflection to zero and if the thickness is such that all photons of energy greater than the band gap are absorbed, the solar energy absorbed can be approximated as follows. The wavelength in nanometers (nm) above which negligible absorption occurs is (1234 nm eV)/1.1 eV = 1122 nm. Looking at the above table and Figure 9, one can estimate that about 80% of the sunlight spectrum would be absorbed, i.e., 80% x 104 W/ft^2

---

10. G. Abetti, "The Sun," Faber and Faber, London, 1963, p. 273, translated by; J. B. Sidgwick.

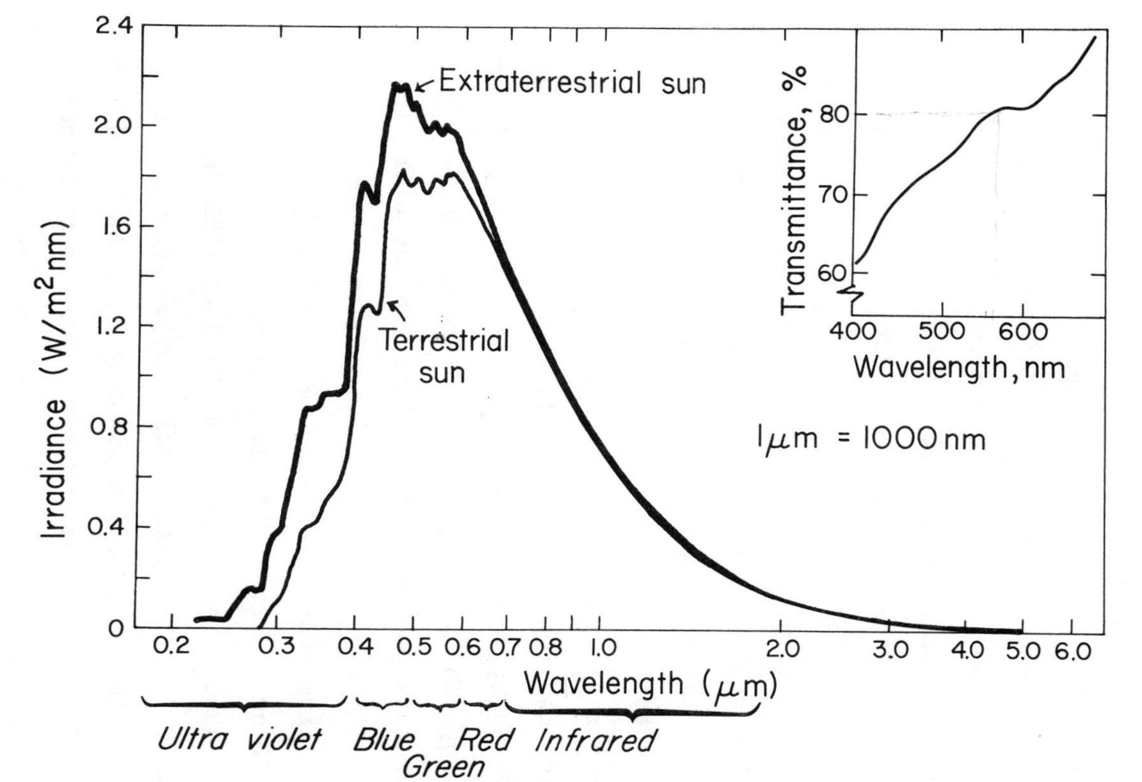

Figure 9. Spectrum of Sunlight and Modulation by the Atmosphere. (Derived from Refs. 8, 9, and 10)

or 83 W/ft^2. Not all of this energy is susceptible to production of electron-hole pairs, however, because the energy in a photon in excess of the band gap goes into other excitation processes which heat the semiconductor. For example, the photons of 561 nm wavelength have an energy of (1234 nm eV)/561 nm = 2.2 eV, just double the energy needed to produce a photocarrier pair. Hence, 50% of the energy in the spectrum at 561 nm will go into heating and not into production of electricity.

The radiant flux reaching a solar cell is subject to the variations of weather and angle between the surface of the cell and the sun. At angles other than normal to the sun, the flux is reduced moderately. Of course reflections from clouds or particulate air pollution can reduce the intensity considerably. Absorption by typical atmospheric gases do not influence the energy reaching the ground to a great degree, but do alter the spectrum, particularly in the ultraviolet and infrared regions (Figure 9). Ozone absorbs strongly from 200-350 nm and some at 602, 4700, 9600, and 14200 nm. Carbon dioxide absorbs strongly at 2700 and 4300 nm and less from 13000 to 17000 nm. Water vapor is the chief barrier to atmospheric transparency in the infrared, absorbing strongly at 1400, 1850, 2700, and 6300 nm and from 14000 nm to longer wavelengths. Besides these strong absorptions in the relatively less energy-rich regions of the solar spectrum, well resolved spectra[5,6] reveal small, narrow absorption bands in the visible region by water vapor and oxygen. These bands are at 594, 652, and 723 nm for water and 629, 688, and 762 nm for oxygen. The coarse spectrum in the visible region where most of the energy is concentrated is nearly constant from location to location, and the direct, focusable radiation (about 60% of the total) has nearly the same spectrum as reflected sky radiation.

As noted in the second chapter, although the irradiance in a plane normal to the overhead sun on a clear day is about 100 W/ft^2 the energy falling at ground level averaged over the U.S. and averaged over a one year period (day and night, all seasons) is 17 W/ft^2. The distribution of this averaged irradiance is shown in Figure 5. One would expect it to vary from zero to slightly greater than 100 W/ft^2 depending upon altitude, time of day and year, and location.

Semiconductor Absorbance - As briefly illustrated above, the absorption properties of photovoltaic materials considerably influence how much of the incoming solar energy is converted to electricity. This is true not only because of the fraction of the solar energy which is absorbed, but also because of where in the semiconductor the majority of photons are absorbed with respect to the surface and p-n junction, and how much energy is converted to heat instead of electricity.

Referring to Figure 7, only those photons with energy greater than the band gap are strongly absorbed. For silicon this means that solar photons of greater energy than 1.1 eV (wavelengths less than 1122 nm) are absorbed in a relatively short distance into the material. Photons of longer wavelengths are not absorbed readily and hence the solar photon spectrum at wavelengths greater than 1122 nm is not appreciably intercepted by silicon. Of the spectrum which is absorbed, 1.1 eV per photon goes to form a hole-electron pair while energy in excess of 1.1 eV/photon goes to heating the crystal.

The absorption spectra for Si and several other semiconductors used in photovoltaic devices[11] are shown in Figure 10. The absorption edge corresponds to the band gap energy, that is, for Si 1100 nm corresponds to a band gap of 1.12 eV and for CdS 520 nm corresponds to a band gap of 2.37 eV. Using the equation for the fraction of incident light of a given wavelength absorbed as a function of distance into the crystal; fraction absorbed = $1-\exp(-\alpha d)$, where α is the absorption constant and d is the distance into the crystal, one can calculate that for 800 nm photons one half would be absorbed within 7 μm of the surface of Si and CdTe, and within 1 μm for GaAs at 300°K. The

11. P. Rappaport and J. Wysocki, Acta Electronica, 5, 364 (1961).

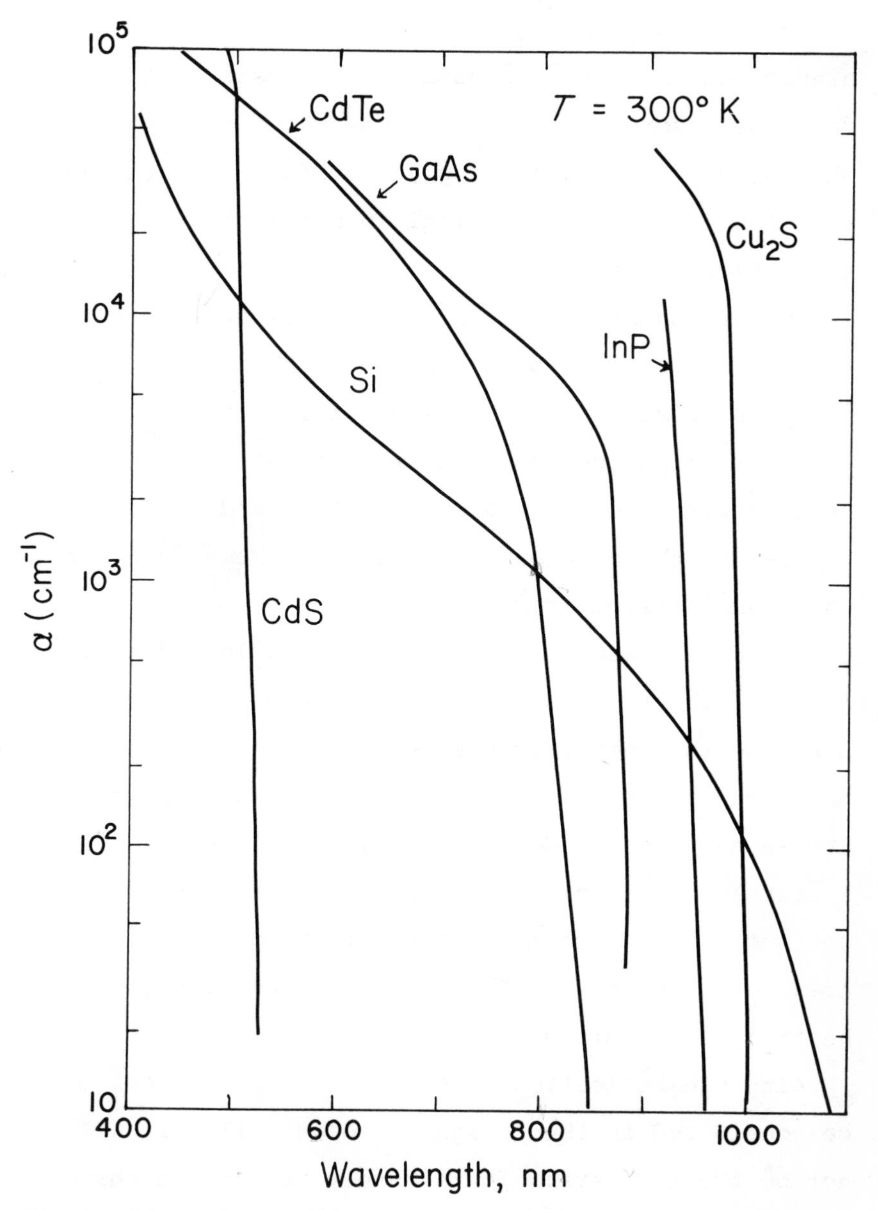

Figure 10. Absorption Spectra of Various Photovoltaic materials. (Printed with permission of G. Pouvesle, Ed., Acta Electronica, Ref. 11.)

higher the absorption constant the closer to the surface the absorption occurs, hence the p-n junction should be located accordingly. Surface effects and resistances are more critical in materials with higher absorption constants.

Comparison of Figures 9 and 10 indicates that semiconductors of lower band gap energy absorb a larger fraction of solar energy. The total flux of extraterrestrial photons and the fraction of incident photons with energy greater than the band gap decreases as the band gap increases as shown in Figure 11 which was derived from references 12 and 13. As the band gap decreases (absorption goes to longer wavelengths) the fraction of sunlight absorbed increases. Likewise the energy losses due to heating by absorbed photons with energy in excess of the band gap increase as band gap decreases. Thus on the basis of spectral considerations alone, there would be an optimum band gap where the sum of the gain from increased solar energy absorption and the loss from heating is maximized.

Although reflection of sunlight from surfaces must be considered in the design of solar cells, it will not be treated here. It is merely recognized that coatings can be used to make losses from reflection negligible.

---

12. J. Wysocki, Solar Energy, 6, 104 (1962).

13. R. Halsted, J. Appl. Phys., 28, 1131 (1957).

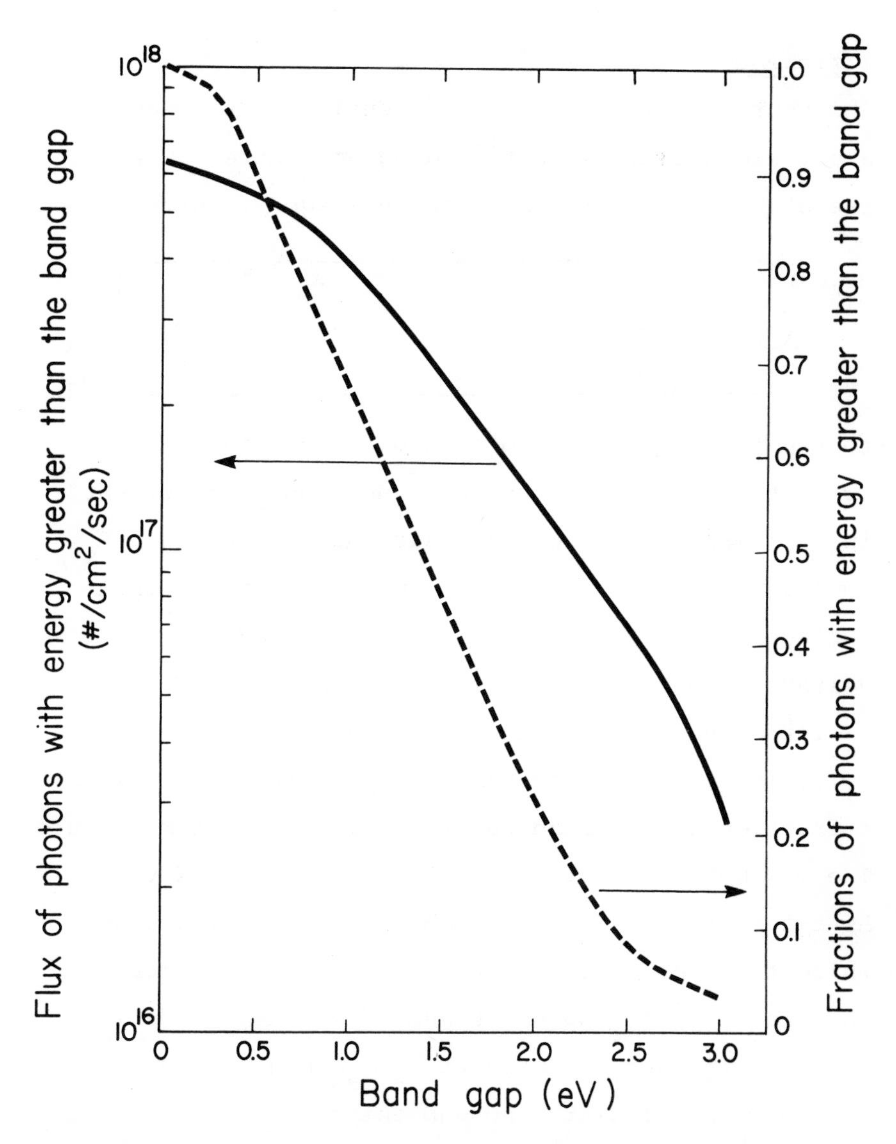

Figure 11. Flux or Fraction of Solar Photons with Energy Greater than a Given Value. (Derived from Refs. 12 and 13)

Efficiency

The photovoltaic cell can be visualized as an equivalent electrical circuit[14] diagramed here. The resistances R, currents I, and voltage V, correspond

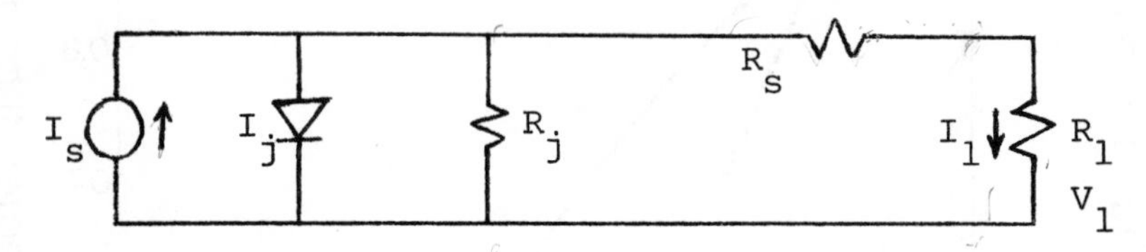

to the following: I_s=total separated charge, I_j= current leaked across the p-n junction, I_l= current in the load, R_j= junction resistance to leakage, R_s= series resistance of the cell mainly from the thin surface layer, R_l= resistance of the load, and V_l= voltage across the load and hence the cell.

The open circuit voltage of the cell, i.e., when R_l is infinite, is the maximum V_{max}. This is always a fraction of the band gap voltage at room temperature. The maximum power of an ideal solar cell is $I_s V_{max}$. As a function of increasing band gap V_{max} increases while the number of photons in the solar spectrum capable of producing holes and electrons, hence I_s, decreases. Then the product $I_s V_{max}$ exhibits a maximum for a particular band gap energy.

Although I_s is fairly constant with respect to temperature (except for minor effects of band gap shifts), the junction resistance to leakage decreases

---

14. J. Loferski, J. Appl. Phys., 27, 777 (1956).

and current leaked, I_j, increases as temperature increases. Hence the maximum voltage possible across the cell decreases with increasing temperature and the power output, I·V, decreases relative to the input of solar energy. Because of the complex manner in which V_{max} varies with temperature, the band gap at which the efficiency is optimum shifts to higher values as temperature increases.[15]

Figure 12 is an illustration of the variation of theoretical efficiency of solar cells made from semiconductors of increasing band gap energy. It is typical of various theoretical treatments which agree in general but not in complete detail.[13,14,15,16,17] The efficiencies illustrated do not take into consideration any losses from cell resistance, photocarrier collection inefficiencies, reflections, transmission, or nonideal junctions. It is evident that the temperature at which a photovoltaic device operates considerably affects the efficiency of energy conversion and that the optimum band gap for maximum efficiency shifts upward with increasing temperature. The slanted, nearly vertical lines represent the shifts in band gap for a given semiconductor as a

---

15. J. Wysocki and P. Rappaport, J. Appl. Phys., 31, 571 (1960).

16. P. Rappaport, RCA Rev., 20, 373 (1959).

17. W. Shockley and H. Wueisser, J. Appl. Phys., 32, 510 (1961).

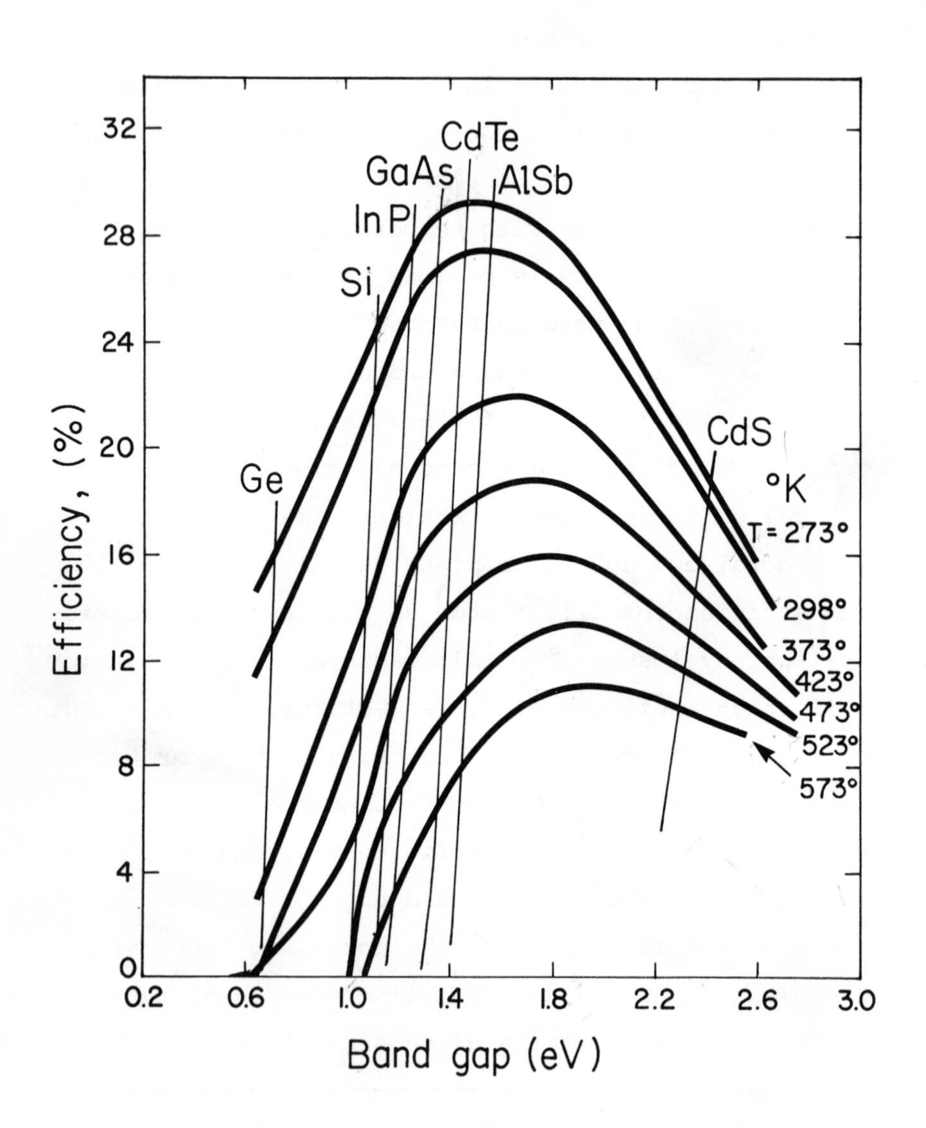

Figure 12. Maximum Theoretical Efficiency of Photovoltaic Converters as a Function of Band Gap Energy.

function of temperature. At room temperature (298°K) the maximum predicted efficiency is about 27% for a material of band gap about 1.4-1.5 eV such as CdTe. The efficiency for CdTe is about halved when the temperature is raised to 200°C.

The maximum power output of a solar cell is $V_{mp}I_{mp}$ which is usually somewhat less than the theoretical maximum $V_{max}I_s$. As illustrated in Figure 13, the open circuit voltage and the current across an experimental Si cell increases as light intensity increases.[16] And at a given light intensity by varying the load resistance one can derive a voltage versus current curve[18] from which V_{mp} and I_{mp} can be found. Comparing $V_{mp}I_{mp}$ with $V_{max}I_s$ shows that the former is about three fourths the magnitude of the latter and that V_{mp} and I_{mp} are less than V_{max} and I_s, respectively.

Power and efficiency is a function of the variation of V_{max}, hence V_{mp}, and I_{mp} with temperature. These are illustrated in Figure 14 which was derived from reference 15. The percentage decrease of V_{max} with increasing temperature is greater than that for I_{mp}, hence the decrease in efficiency with increasing temperature is predominantly attributable to the voltage variation. The temperature effect on efficiency

18. R. Gold, "Current Status of GaAs Solar Cells," Transcript of Photovoltaic Specialists Conference, Vol. 1, Photovoltaic Materials, Devices and Radiation Damage Effects, DDC No. AD412819, July, 1963.

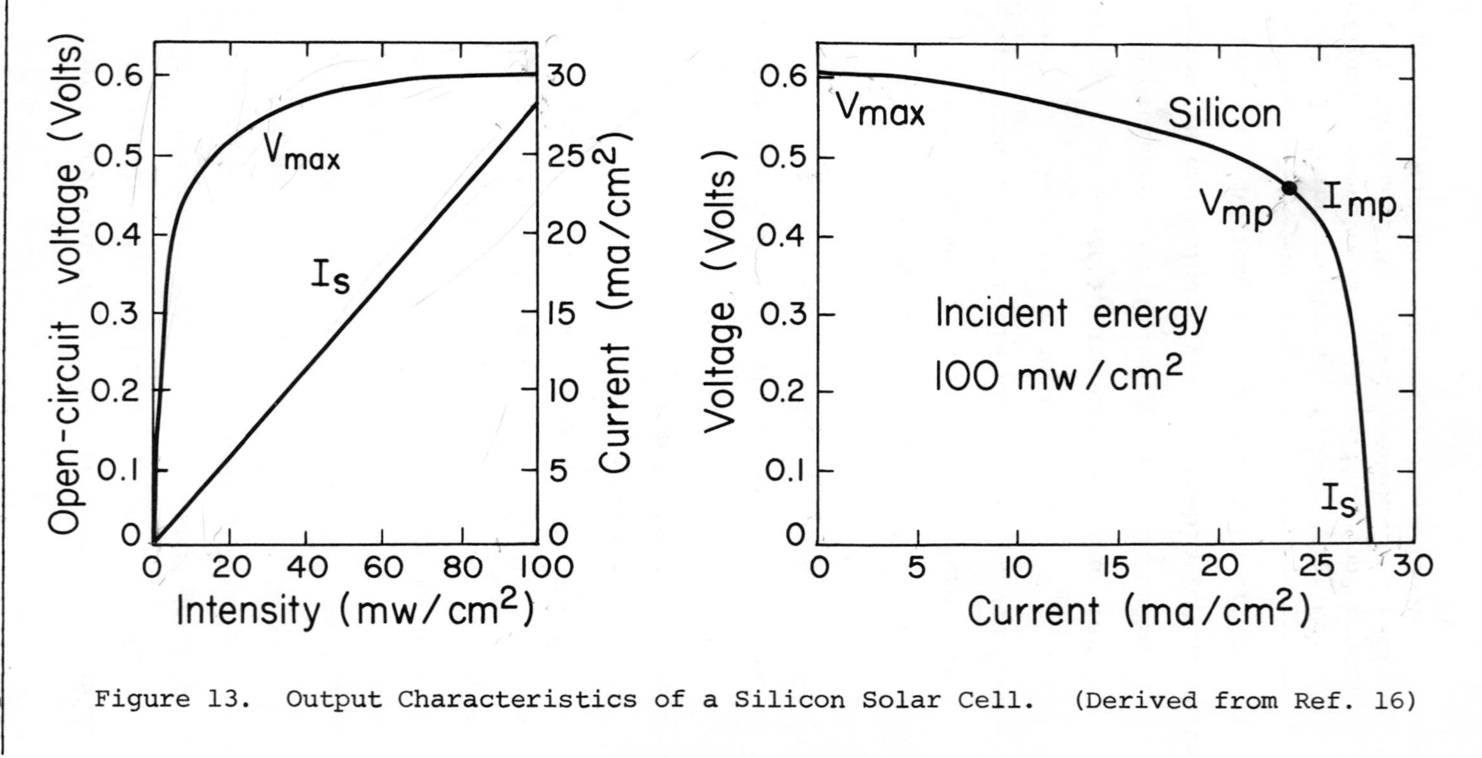

Figure 13. Output Characteristics of a Silicon Solar Cell. (Derived from Ref. 16)

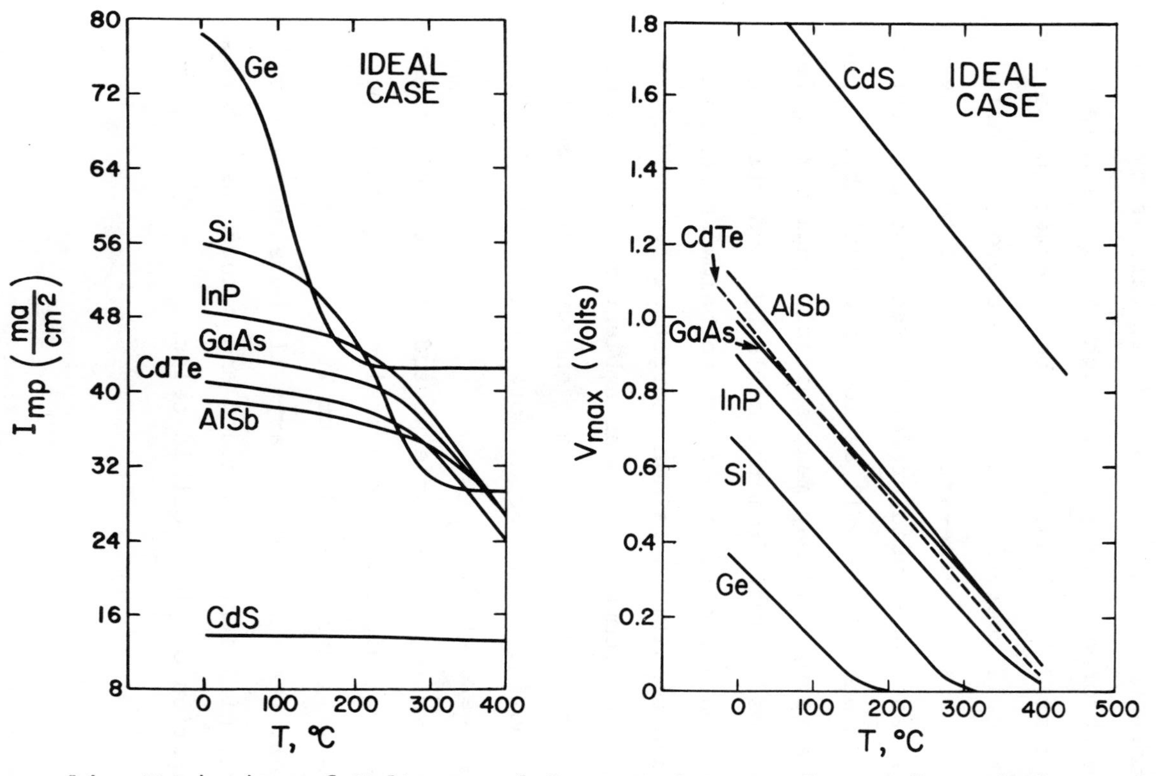

Figure 14. Variation of Voltage and Current Outputs from Selected Photovoltaic Materials as a Function of Temperature. (Derived from Ref. 15)

is less for semiconductors of higher band gap. Also, the higher the band gap, the smaller is the fraction of absorbed photon energy which goes to heating the semiconductor. In the case of Si, this is over 40% of the absorbed energy. Although Si may be more efficient at 25°C, ideal CdS may be a better choice for a solar cell operating at 100°C or above, depending upon the state-of-the-art in solar cell technology. In any case, one would prefer to keep the temperature of the photovoltaic cell as low as practicable. For large arrays, this may mean suitable engineering for heat transfer away from the cells by conduction and convection. Cooling the cells to below ambient temperature conditions is not energetically thrifty. Keeping the temperature near ambient may not be too difficult in the very diffuse energy flux from the sun. Cooling would become more critical if focusing devices were used.

In summary, it is not expected that the efficiency of photovoltaic energy conversion at ambient temperatures will ever exceed about 25%. In the next chapter, the state-of-the-art will be examined with regard to efficiencies obtained and methods used for preparation of photovoltaic cells.

Chapter 4

STATE-OF-THE-ART IN PHOTOVOLTAIC CONVERSION TECHNOLOGY

There is no dearth of problems to be overcome before photovoltaic energy conversion competes economically with other means of supplying electricity on a broad basis. Technological problems include the proper control of manufacturing processes to yield high junction potentials, high collection efficiencies, very pure defect-free cells, and electrode contacts with low resistance. Much engineering refinement is needed for building large arrays with many cells-to-substrate contacts and cooling capabilities. And most substantially, methods for fabricating semiconductors of adequate quality at a much lower cost must be found. Because the manufacture of large single crystals (2-5 cm^2) of the requisite purity is an expensive process, emphasis is going toward thin films. These photovoltaic films are more polycrystalline and susceptible to inefficiencies, but they can be mass produced much more readily.[1]

If the aforementioned problems were solved, another technological question would have to be answered before photovoltaic energy conversion could reach its highest potential. How can the electrical energy

1. M. Altman, "Elements of Solid State Energy Conversion," Van Nostrand Reinhold Co., New York, 1969, pp. 240-263.

produced in the photovoltaic process be stored in an energy-dense, conveniently redistributable way with high efficiency? Most electricity is used on-demand, which does not always coincide with sunshine. Considerable research and development will be required for more energy-rich, inexpensive chemical storage cells, efficient electrolysis to form hydrogen (H_2) and oxygen (O_2) which can then be used in efficient fuel cells to regenerate electricity, or other forms of energy storage such as flywheels.

Silicon Cells

Practically all of the power used by spacecraft is provided by photovoltaic devices made from silicon. Silicon is the second most abundant element in the earth's crust (as SiO_2) and is produced[2] in the U.S. in metallurgical purity for $600/ton. The cost of this relatively impure Si raw material in a typical one square foot solar array would be four cents. The actual cost of that array is 200-8000 dollars at present.[3] The annual U.S. production, devoted almost entirely to the space program, yields about 50-70 kW

---

2. C. Currin, K. Ling, E. Ralph, W. Smith, and R. Stirn, "Feasibility of Low Cost Silicon Solar Cells," Proceedings of the 9th IEEE Photovoltaic Specialists Conference, Maryland, May 1972.

3. See reference 2, Chapter 2.

of electricity.[4] This very high cost and low production stems from the complex manner in which silicon solar cells are produced.

Because of the need for extremely high purity and single crystals to prevent recombination and loss of photocarriers, silicon is usually grown in ingots about 3 cm in diameter which are drawn slowly out of a hot melt of pure Si. Wafers of Si about 200-500 μm thick are cut from the ingot in a delicate fashion so as to prevent defect formations. These wafers of Si, or doped Si, are then subjected to etching, sandblasting, electron bombardment or other techniques for removing surface defects. They are next heated in the presence of an appropriate chemical which allows diffusion of a dopant into the surface to yield a p-n junction. The wafer may then be subjected to more etching to adjust the surface-layer-to-junction distance and to coatings of various types for electrode contacts or reflection reduction. After electrodes are attached (often by elaborate and tedious procedures) and the cell is protected by lacquer or perhaps a fused-on glass covering, it is ready for use. This description of Si cell preparation is much too simplistic to convey the true art and intricacy of the processes used. For high efficiency cells much more elaborate crystal growing and regrowing

---

4. M. Wolf, "Historical Development of Solar Cells," 25th Power Sources Conference, Atlantic City, May 1972.

techniques, complicated profile doping, and intricate electrode contact procedures are utilized.

To make large scale terrestrial use of Si solar cells economically feasible the cost will have to decrease by about 2-3 orders of magnitude. Part of this is expected through automation and combining of steps. However, new processes are needed which lend themselves to mass production. Many ideas have been advanced and experimental work is in progress.
M. E. Paradise[5] has patented a process for using small beads of Si about 2 mm in diameter which are coated in a plastic layer which is then subjected to etching, doping, and contacting processes. This promises a large area photovoltaic converter at lower cost, but decreased efficiency relative to the wafers. Of perhaps higher potential are techniques for growing ribbons or sheets of large-crystal Si by rolling Si,[6] casting sheets and recrystallizing through heated or molten zones,[7] dendritic growth,[8] and edge-defined

---

5. M. E. Paradise, U.S. Patent 2,904,613; September 15, 1959.

6. W. R. Cherry, Proceedings of the 13th Annual Power Sources Conference, Atlantic City, May 1959, pp. 62-66.

7. Ryco Laboratories, Final Report No. AFCRL-66-134, 1965.

8. R. Riel, "Large Area Solar Cells Prepared on Silicon Sheet," Proceedings of the 17th Annual Power Sources Conference, Atlantic City, May 1963.

film growth.[2] It has long been known that Si cells need not be over about 100-150 μm thick for high efficiency cells,[9] hence it is probable that thin film techniques will eventually reduce the cost of manufacture of these cells. One can envision a continuous process of thin film formation, heating, ion implantation, annealing, and vapor deposition of coating and contacts.

The average efficiency of present production silicon cells is 13-14% for terrestrial applications with 15-16% reported on experimental cells.[10] This is possible by close attention to cell design to maximize collection efficiency and minimize resistance losses. The graded-base cell, which extends the junction electric field region into the bulk by a graded concentration dopant, has helped increase the collection efficiency.[11] It has also been enhanced by precise control of the surface-to-junction distance to minimize surface recombination and resistance losses. A

---

9. R. Crabb, "Status Report on Thin Silicon Solar Cells for Large Flexible Arrays," Solar Cells, Gordon and Breach Science Publishers, N.Y., 1971, pp. 35-50.

10. J. Lindmayer, Proceedings of the 9th IEEE Photovoltaic Specialists Conference, Maryland, May 1972.

11. S. Kaye, "Drift Field Solar Cells," Transcript of the Photovoltaic Specialists Conference, Vol. 1, Photovoltaic Materials, Devices and Radiation Damage Effects, DDC No. AD412819, Sec A-6 July 1963.

recent study[12] showed that if it is possible to obtain photocells with the p-n junction depth less than 0.5 μm with reliable contacts, it is as good as a graded-base cell. However, technology of making thin layers and good contacts is difficult. One technique provides for monitoring the output current under constant irradiation while the surface is being etched.[13] For decreasing the surface resistance while maximizing collection efficiency, Huth[14] has patented a cell whose surface layer thickness increases as the electrical contact is approached. Precise control of surface layer doping has been achieved which allows the highest surface conductivity without a degenerate layer which is accompanied by unwanted absorption.[15] The optimal doping yields the highest concentration of dopant possible to maximize conductivity and junction potential without reaching the state where unwanted light absorption and leakage through the junction predominates. Through ion implantation it is possible to precisely define the dopant concentration profile.[16]

---

12. N. Gavrilova, Applied Solar Energy, 8, No. 5-6, 68 (1973).

13. C. Beauzee, U.S. Patent 3,261,074; July 19, 1966.

14. J. Huth, U.S. Patent 2,993,945; July 25, 1961.

15. S. Miller, U.S. Patent 3,081,370; March 12, 1963.

16. V. Reddi and J. Sansbury, J. Appl. Phys., 44, 2951 (1973).

New doping techniques along with higher purity, defect-free silicon may aid in increasing the junction potential in future cells.

Another significant factor which has allowed increased efficiency is advancement of the thin coating technology on the face of the photocell. Various coatings such as silicon monoxide,[17] titanium dioxide, zinc sulfide, and cerium oxide[18] serve to reduce reflection from 30-60% to 3-6%.

An ad hoc panel on solar cell efficiency under the sponsorship of the National Research Council[19] recently reported the outlook for improved efficiency of silicon solar cells has not increased significantly for the last five years. This appears to be due not to fundamental limitations, but to a cessation of research and development in this area which with presently utilized technology does not offer significant profits to manufacturers. The market through the 1960s was for about 1-2 million cells per year, at an average of $15 per 2 x 2 cm fully operational cell.

17. P. Iles and B. Ross, U.S. Patent 3,361,594, January 2, 1968.

18. A. Atzei, J. Capart, R. Crabb, K. Heffels and G. Seibert, "Improved Antireflection Coatings on Silicon Solar Cells," Solar Cells, Gordon and Breach Science Publishers, New York, 1971, pp. 349-362.

19. Ad Hoc Panel on Solar Cell Efficiency, P. Rappaport, Chairman, "Solar Cells; Outlook for Improved Efficiency," National Academy of Sciences, Washington, D. C., 1972.

In the 1970s the demand is expected to be about the same at the same price. With the appropriate incentive and about 250 person-years of research it is predicted that an efficiency of 22% for terrestrial applications could be obtained. This figure is based partially on an analysis of technical factors by M. J. Wolf.[20] His analysis of the present conversion factors is shown schematically in Figure 15. Seventy seven percent of the incident solar energy is absorbed in a state-of-the-art cell. Forty three percent of that which is absorbed is lost through heating of the crystal because of photon energy in excess of the band gap. The "voltage factor" which is the ratio of V_{max} to the band gap energy, E_g, is 0.62 for cells designed for terrestrial use (0.49 in cells designed for space use). The "curve factor" of 0.82 refers to the theoretically ideal ratio $V_{mp}I_{mp}/V_{max}I_s$. The additional curve factor A of 0.90 is caused by nonideal deviations in $V_{mp}I_{mp}/V_{max}I_s$ from increased current flow through the junction of the cell at the expense of current through the load. The combination of curve factor effects was illustrated in Figure 13. The efficiency of collecting the holes and electrons is about 78%, and losses from the surface resistance and reflections amount to 3% each. Thus, the output efficiency is about 14% for state-of-the-art terrestrially

20. M. J. Wolf, "The Fundamentals of Improved Silicon Solar-Cell Performance," Chapter 4 of reference 19.

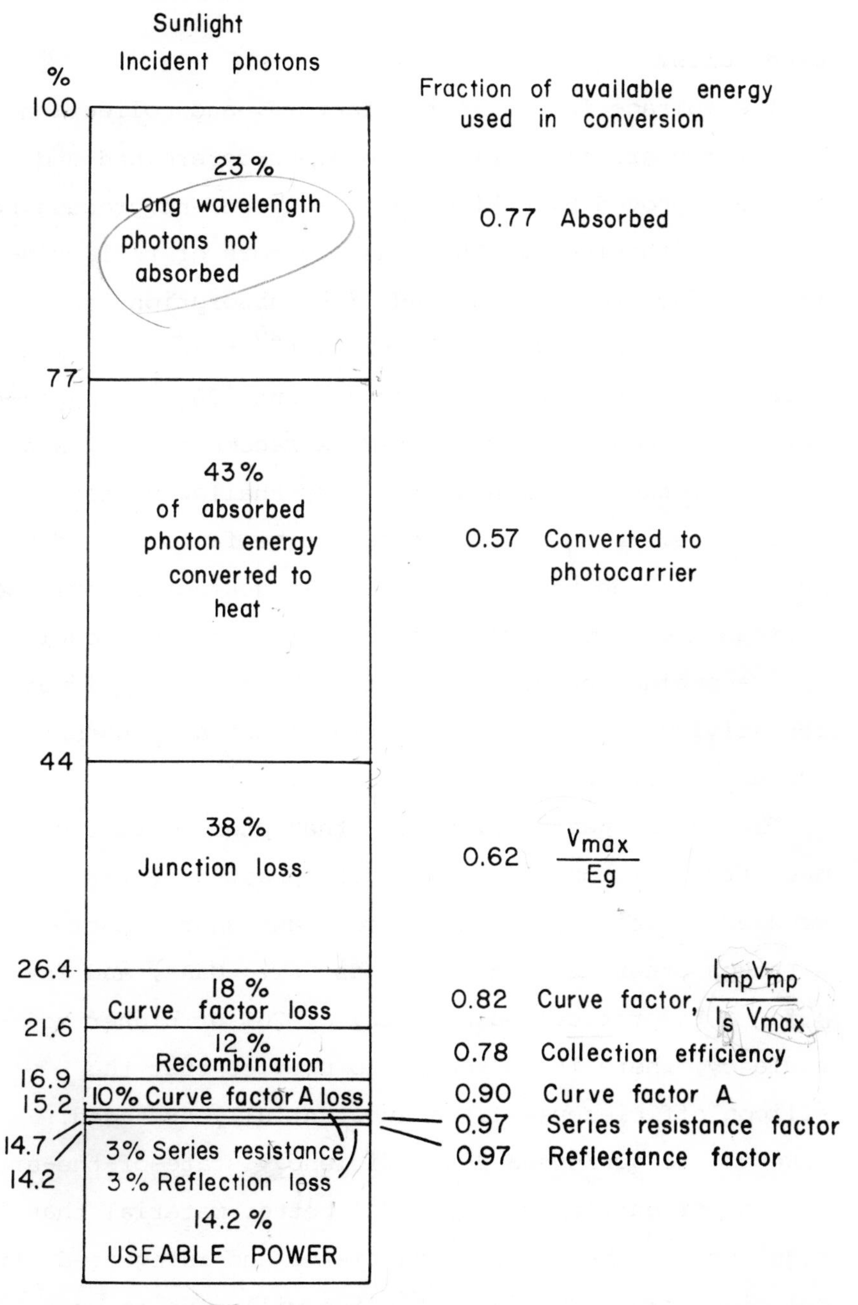

Figure 15. Photovoltaic Conversion Factors in Silicon Cells. (Derived from Ref. 20)

used cells.

The voltage factor, curve factors, and collection efficiency are materials and process-determined and may be improved by making purer silicon and processing it more effectively. The other factors of reflectivity, surface resistance, and light absorption cannot be improved readily. It is thought[20] that the collection efficiency can be improved about 20% by using better silicon which would reduce recombination, and by making the junction depths very shallow on the order of 0.1-0.2 μm. The total curve factor could be increased by about 15% and the voltage factor could be increased by about 40% by close attention to reducing the recombination center density and achieving lower resistivity. Thus the predicted technically possible conversion efficiency is above 20%.

The ad hoc panel believes[19] that photovoltaic cells made from silicon offer the most opportunity for increased efficiency at lowest cost and shortest time. Although other materials such as CdTe, GaAs, and InP offer a theoretical possibility of 20% or higher efficiency, there is no reason to predict that the silicon efficiency will be substantially exceeded. However, to achieve a 20% efficiency, state-of-the-art silicon production must yield a better material than is required for the present transistor and integrated circuit industry. Special attention will have to be devoted to the material quality requirements.

The key problems to be overcome before silicon solar cells reach commercial viability for massive terrestrial use are:

1. The costs of production must be reduced[3] from the $200 per ft^2 expected for the 1970s to $0.50—$3.50 per ft^2.
2. Mass production high technology processes must be devised.
3. A useful lifetime of 20 years should be attained.
4. Efficiency should be increased to near the theoretical limit of about 25% to compete with other methods of using solar energy to provide electricity.

Of these problems, the third appears solvable based upon existing data.[3] Silicon cells used in space applications are expected to have this lifetime with no degradation. However, in the terrestrial environment they are subjected to moisture, sulfur oxides, nitrous oxides, dust, winds, ice, pollution of various kinds,...etc. Good protection with full transmission of light will be necessary. The fourth problem was discussed above, and the first problem depends very heavily on solving the problem listed in number two. With present efforts, predominantly in academic institutions and without a great deal of funding, these problems are not likely to be overcome in the next 10-15 years.

Cadmium Sulfide-Cuprous Sulfide

Cadmium sulfide as used in photocells is generally an n-type semiconductor perhaps doped with impurities such as $InCl_3$, $GaCl_3$, BCl_3 or excess Cd to provide increased conductivity.[21] The CdS base material is usually treated with an aqueous cuprous solution which causes some conversion of the CdS to Cu_2S and various intermediate compositions to cause an n-p junction. The base CdS is contacted by such metals as indium, tin, or zinc while the thin Cu_2S surface layer is contacted by copper or gold. Because of the irregular composition of the junction, its unusual absorption and electrical properties, and deviations from the better understood solid state physics of silicon-type devices there have been many conflicting theories of how the cell operates and what efficiency one can expect from it.[22] Because the temperature sensitivity and the spectral sensitivity of these cells do not conform to what is expected from the absorption (Figure 10) and the theoretical calculation for CdS (Figure 14), it is recognized that Cu_2S and junction

---

21. D. A. Gorski, U.S. Patent 3,186,874; June 1, 1965.

22. A. VanAerschot, J. Capart, K. David, M. Fabbricotti, K. Heffels, J. Loferski, and K. Reinhartz, "The Photovoltaic Effect in the Cu-Cd-S System," Conference Record of the 7th Photovoltaic Specialists Conference, Pasadena, Calif., November 1968, p. 22.

material play the major role in production of photo-carriers.[23]

The energy band diagram which is believed to represent the heterojunction Cu_2S-CdS is illustrated in Figure 16, derived from reference 23. Absorption of light in the Cu_2S which has a band gap of 1.2 eV causes the photocell to be photosensitive to light from the ultraviolet to 1000 nm. Absorption may also occur in the junction itself where an intermediate material composition is present. Flow of photoelectrons from the p-type Cu_2S to the n-type CdS changes the representation of energy levels when the cell is illuminated as shown.

Because both CdS and Cu_2S are strongly absorbing materials compared with Si (see Figure 10), light may be absorbed in a very short distance. This allows these cells to be much thinner than silicon cells. Thus material can be conserved and thin films can be used. However, the junction needs to be close to the surface and high surface layer losses may be encountered. Although large single crystals of CdS can be melt-grown, much like silicon, and then treated with p-typing cuprous ions and used as a solar cell, much easier processes can be used. A powder of CdS may be sprinkled or pressed onto a supporting substrate such

---

23. L. Shiozawa, G. Sullivan, and F. Augustine, "The Mechanism of the Photovoltaic Effect in High Efficiency CdS Thin-Film Solar Cells," ibid., p. 39.

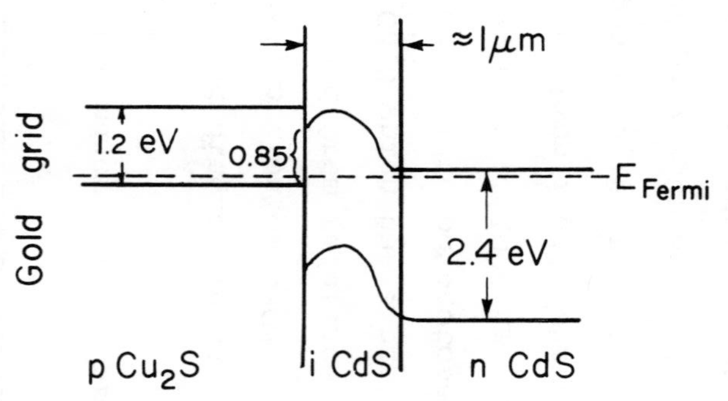

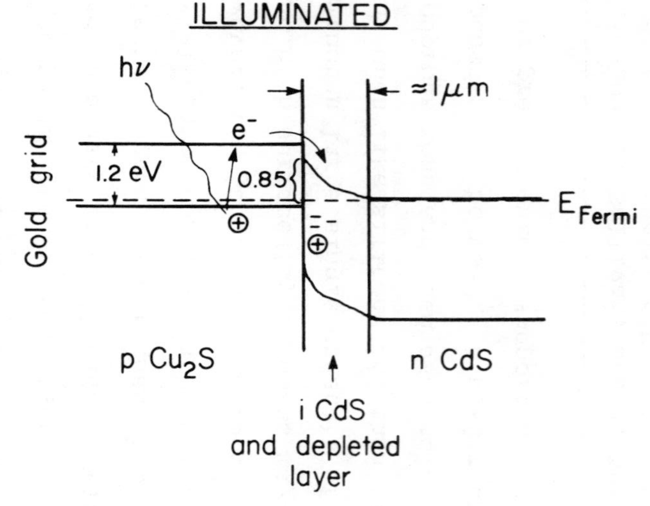

Figure 16. Energy Band Structure in the Dark and Illuminated CdS-Cu_2S Heterojunction. (Printed with permission of the Institute of Electrical and Electronic Engineers, Inc., Ref. 23.)

as glass or a metallic contact like zinc, sintered at a high temperature in an inert atmosphere, and soaked in a cuprous solution. After a surface contact is applied, such a rudimentary cell may achieve a 3% efficiency of solar energy conversion.[24]

By far the most utilized technique for preparing CdS solar cells is by vapor deposition. Typically, the CdS is vacuum deposited onto a Kapton, polyimide flexible film about 25-100 μm thick which had been previously coated with silver plus polyimide varnish and plated with zinc. The CdS layer, which is about 5-50 μm thick for optimal performance,[25] is then bathed in an acidic solution of cuprous chloride or otherwise treated[26] by sputtering or flash evaporation of Cu_2S to an optimum thickness of 3.5 μm or less.[27]

---

24. V. Komashchenko, A. Marchenko, and G. Fedorus, Poluprovodnikovoya Tekhnika; Makroelectronika, No. 4, 112 (1972). AD-756594, NTIS, U.S. Dept. of Commerce, Springfield, Va.

25. I. Abrahamsohn, U.S. Patent 3,376,163; April 2, 1968.

26. J. David, S. Martinuzzi, F. Cabane-Brouty, J. Sorbier, J. Mathieu, J. Roman, and J. Bretzner, "Structure of CdS-Su_2S Heterojunction Layers," Solar Cells, Gordon and Breach Science Publishers, New York, 1971, pp. 81-94.

27. S. Yu Pavelets and G. A. Fedorus, Geliotekhnika, 7, No. 3, 3 (1971). Applied Solar Energy, 7, No. 3-4, 1 (1973).

A metal such as gold is then evaporated onto the Cu_2S layer to provide an electrical contact. It may be thin enough to be partially transparent. Or it can be thick enough to be masking and photoresist techniques can be used to make a grid of metal contacts. The surface is then covered with a transparent adhesive which adheres a 25-100 μm thick Kapton, polyimide protective cover. A diagram of a typical structure is shown in Figure 17. The resulting solar cell is polycrystalline, flexible, and protected by front and back layers of polyimide.

Because the technique described can be done with automatic equipment on a massive scale the price per square foot of future commercially available cells is expected to be much lower than for silicon cell arrays. A cost of $0.50 per ft^2 is a reasonable aim.[3] Because the CdS and Cu_2S layers are so thin, the materials resources are not expected to be a constraint.

The factors which appear to limit the general use of these cells are their relatively low efficiency and stability. The V_{max} for a typical cell is 0.46-0.48 V at room temperature.[28] However, experimental cells have been reported[27] with V_{max} as high as 0.57 V. The curve factor is rather low, and the maximum power voltage is about 0.33-0.40 V. Typical efficiencies

28. A. G. Standley, "Present Status of Cadmium Sulfide Thin Film Solar Cells," Technical Note 1967-52, Lincoln Laboratory, M.I.T. Lexington, Mass. 1967.

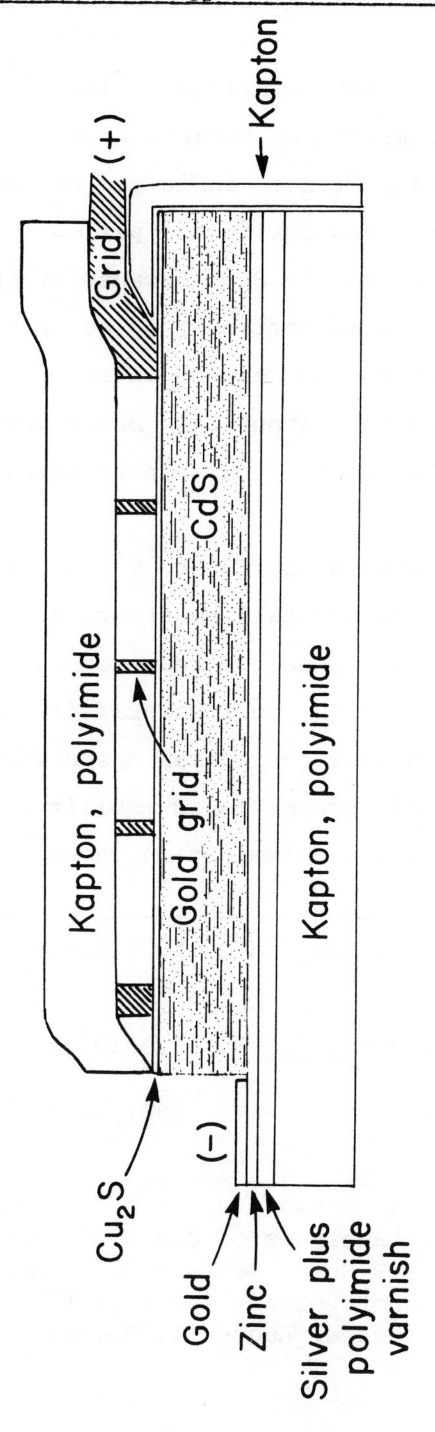

Figure 17. Typical Construction of a CdS-Cu_2S Photovoltaic Converter

are 6-9% for recent thin film cells.[29,30,31] A recent report[32] of research at Societe Anonyme de Telecommunications, Paris, France, indicates an increase in efficiency achieved during the period 1967-1970 from less than 1% to nearly 8%. Because of the complexity of the cells, it is difficult to estimate the efficiency which may be reached ultimately. The highest estimation would be about 24%, but a more probable limit under the best circumstances foreseeable is 18%.[27]

The main reasons for CdS-Cu_2S cell instability are: (1) the polycrystalline structure presents a high effective surface area for reactions with chemical contaminants, and (2) cuprous sulfide has a phase transformation near the normal operating temperature and considerable copper ionic mobility. The combination of heat and illumination causes a decrease in

---

29. A. I. Marchenko and G. A. Fedorus, U.F.Zh., 12, 1392 (1967).

30. I. V. Egorova, F.T.P., 2, 319 (1968).

31. T. Shiata and N. Sato, Gap. G. Appl. Phys., 7, 1348 (1968).

32. W. Palz, J. Besson, J. Fremy, T. Nguyen Duy, and J. Vedel, "Analyses of the Performance and Stability of CdS Solar Cells," Conference Record of the 8th IEEE Photovoltaic Specialist Conference, Seattle, Washington, August 1970, pp. 16-23.

output efficiency.[33] Although lengthy illumination of cells at 25°C to light intensities equivalent to slightly more than direct sunlight causes some degradation of V_{max}, it is reversible by holding the cells in the dark or short-circuiting them.[32] Irreversible degradation occurs when the cells are illuminated and heated to temperatures above 60°C. This is associated with an increase in ionic conductivity and $\gamma \longrightarrow \beta$ phase transition in the region of 80°C.

The degradation of CdS-Cu_2S cells is much more rapid at high humidity and is a strong function of the electrical load. Under open circuit conditions and illumination, the power output may drop to about two thirds its original power in about three days.[34] However, under lighter load conditions, this does not happen as readily. The cause of the power loss is thought to be the degradation of Cu_2S to CuS and Cu.[35]

---

33. A. G. Stanley, "Degradation of CdS Thin Film Solar Cells in Different Environments," Technical Note 1970-33, Lincoln Laboratory, M.I.T., Lexington, Mass. 1970.

34. A. Spakowski and A. Forestieri, "Observations of CdS Solar Cell Stability," Conference Record of the 7th IEEE Photovoltaics Specialists Conference, Pasadena, California, November 1968, p. 155.

35. D. Bernatowicz and H. Brandhorst, Jr., "The Degradation of Cu_2S-CdS Thin Film Solar Cells Under Simulated Orbital Conditions," Conference Record of the 8th IEEE Photovoltaic Specialists Conference, Seattle, Washington, August 1970, pp. 24-29.

This reaction has an electrode potential at 25°C of -0.39 V. When the applied voltage across the surface layer of a $CdS-Cu_2S$ cell exceeds this copper is produced which shorts the cell. Thus at open circuit voltages of 0.47 V, the cell rapidly degrades. Fortunately, the maximum power condition is when the voltage is about 0.33 V, far enough below the threshold that degradation will not occur from this reaction at ambient temperatures and proper selection of load resistances. Other slower failures occur however. One might expect a cell to lose about 10% power output after about a one year exposure to sunlight at 25°C.

In summarizing the problems to be solved before $CdS-Cu_2S$ cells are commercially viable in large markets, one can foresee that:

1. The cost of production[3] should be reduced to about $0.50/ft^2.
2. Presently available technology and novel engineering methods must be utilized on a mass production basis.
3. A useful lifetime of 5-10 years should be achieved, although this could be less for some market areas.
4. Means of protecting the cells from high temperature operation will be needed unless a new junction of greater stability than Cu_2S affords is devised.

5. Increased efficiency, perhaps in the 12% region or greater, would be required to compete with other means of using solar energy to produce electricity. This depends upon the cost of the cells and the value of space used for their deployment.

Other Photovoltaic Materials

Materials which have been investigated for use as solar cells include GaAs, InP, GaP, CdTe, and others. The main incentive for investigating these materials is to achieve a higher efficiency than is theoretically possible with Si or CdS, and to achieve better power output and stability at high operating temperatures. The disadvantages of working with them are their relatively high costs, problems with impurities, and complex crystal growing methods in some cases. Efficiencies of up to 8% have been reported for InP, GaAs, and CdTe.[36] Recently, a special gallium aluminum arsenide photovoltaic cell with 16% efficiency was reported.[37] However, it was very difficult to prepare, and is thought to cost 10-100 times the cost of a silicon cell. A junction of p-$GaAs_xP_{1-x}$—n-GaAs has

36. P. Crossley, G. Noel, and M. Wolf, NASA Report NASW-1427.

37. J. Woodall and H. Hovel, 141st Meeting of the Electrochemical Society, Houston, Texas, May 1972. Appl. Phys. Let., 21, 379 (1972).

been made[38] by high temperature (900°C) liquid epitaxial growth of the p-type material on n-GaAs with a resultant 7% efficiency even though 30% of the incident light was reflected. The cell developed a V_{max} of 0.82 V and absorbed light from 400-900 nm. High voltage photovoltaic cells using tellurium-doped GaAs  have been reported, also,[39] which show a 7% efficiency, higher current density than Si, and absorption from 400-900 nm.

Cadmium telluride, like CdS, is a highly ionic material and as such offers about the same efficiency possibilities in a thin film as in a single crystal. It is typically prepared as a heterojunction cell by vacuum depositing CdS onto a molybdenum contact backing, vapor depositing CdTe, flash evaporating and depositing Cu_2Te, and then evaporating a gold grid.[40] Such cells have a V_{max} of about 0.6 V. Stability problems which plagued these cells in earlier years are now being solved. Typically, a power degradation

---

38. A. Cheban, V. Negreskul, P. Oush, L. Gorchak, G. Unguranu, and V. Smirnov, Geliotekhnika, 8, No. 1, 30 (1972). Applied Solar Energy, 8, No. 1-2, 23 (1973).

39. T. Dorokhina, A. Zaytseva, M. Kagan, A. Polisan, and B. Kholev, Geliotekhnika, 9, No. 2, 6 (1973). Applied Solar Energy, 9, No. 1-2, 50 (1974).

40. J. Lebrun and G. Bessonneau, "New Work on CdTe Thin Film Solar Cells," Solar Cells, Gordon and Breach Science Publishers, New York 1971, pp. 201-206.

of less than 3% occurs after 1000 hours at 70°C.[40] However, sustained illumination causes degradation.[41]

Development of these materials is perhaps five years[42] behind Si and CdS. Relatively little development effort is going toward them, but a fairly steady research interest is continuing. It is improbable that such devices as have been studied will ever be less expensive, and they can have only limited advantages over the Si and CdS materials. However, many materials are yet to be studied and advances in materials science may allow significant breakthroughs in photovoltaic energy conversion technology. Schottky barrier devices deserve much more thorough evaluation for purposes of photovoltaic energy conversion. A Schottky barrier is basically composed of a metal-semiconductor junction which can be simply fabricated by evaporating a metal onto a semiconductor. Because the potential at the barrier (junction) does not extend over a large distance it is advantageous to have a strongly absorbing semiconductor and a nearly transparent thin metal layer through which light is received. Theoretical considerations lead to the conclusion that Schottky barrier solar cells are capable

41. M. Guillien, P. Leitz, G. Marchal, and W. Palz, "Study of Some Photoelectric Properties of CdTe Solar Cells," _ibid_, pp. 207-214.

42. See reference 19.

of a solar conversion efficiency very similar to that of conventional p-n homojunction cells,[43] i.e., about 25% for semiconductors having a band gap between 1.4 and 1.6 eV. It is inappropriate to predict the timing of technological breakthroughs, or even that they will occur, but one cannot ignore the distinct possibilities for novel photovoltaic cells based upon new materials.

Organic materials with semiconductive properties offer some hope for making low-cost easily fabricated photovoltaic cells. Many absorb in the visible region and some cells based upon the Schottky barrier have had a V_{max} of 0.5-1 V. Although these voltages are in the same range as inorganic semiconductors, the currents are very small. Power conversion efficiency of about 10^{-4}% has been reported for a tetracene film sandwiched between aluminum and gold electrodes.[44,45] Magnesium phthalocyanine between aluminum and silver yields an efficiency of 10^{-2}%.[46] Electrodeposited

---

43. D. Pulfrey and R. McOuat, Appl. Phys. Lett., 24, 167 (1974).

44. L. Lyons and O. Newman, Australian J. Chem., 24, 13 (1973).

45. A. Ghosh and T. Feng, J. Appl. Phys., 44, 2781 (1973).

46. A. Ghosh, D. Morel, T. Feng, R. Shaw, and C. Rowe, J. Appl. Phys., 45, 230 (1974).

chlorophyll-a between aluminum and gold or mercury yields up to 10^{-3}% efficiency.[47]

Before the organic semiconductors compete favorably with Si or CdS, improvement of several orders of magnitude will have to be made in respect to lower resistance and higher current flow. Because there is little molecule-to-molecule or atom-to-atom coupling in the molecular organic crystals relative to the ionic inorganic crystals, one might expect that they would have poorer transport of energy, lower photocarrier mobilities, sharper absorption peaks, and more luminescence losses than the inorganic semiconductors.[48] Hence, it appears that organic semiconductors provide little immediate encouragement with regard to efficiency of solar energy conversion. Progress will probably be made by discovery of new materials or new effects as opposed to development of known material and scientific phenomena.[42]

Possibilities for Technological Advancement and Cost Reduction

Concentrating only on the Si and CdS-Cu_2S materials,

---

47. C. Tong and A. Albrecht, to be published, *J. Chem. Phys.*, 1975.

48. H. J. Queisser, "Theoretical Efficiency Considerations for Photovoltaic Energy Converters," Chapter 3 of reference 19.

it is possible to envision ways in which manufacturing technology can be utilized to reduce the cost of photovoltaic energy conversion and provide higher production capability. For the CdS-Cu_2S cell one can visualize a continuous process of zinc plating on a plastic base which has contact electrodes attached, CdS deposition, Cu_2S formation by dipping, vacuum deposition of a metal grid, and application of a protective covering with metal contacts as schematically described in Figure 18. Present day technology provides commercially available vacuum deposition units with inlet and outlet ports for continuous flow of material through the chamber. Such a facility operating at 100 ft^2 per minute and 20% downtime could provide over 40 million square feet of photovoltaic coverters per year. At 7% efficiency and a U.S. annual average of 17 W/ft^2 solar illumination on a horizontal surface, this could provide an average of 50 megaWatts of direct current electricity. If the replacement time were 5 years, such a manufacturing facility could provide for an averaged 250 MW power station with continuous replacement of 5 year-old partially degraded arrays.

At the present stability of CdS-Cu_2S converters it is difficult to envision using them in arrays which allow inexpensive concentration of sunlight and concomitant cost reduction. If the light incident on the material were of much greater intensity than direct

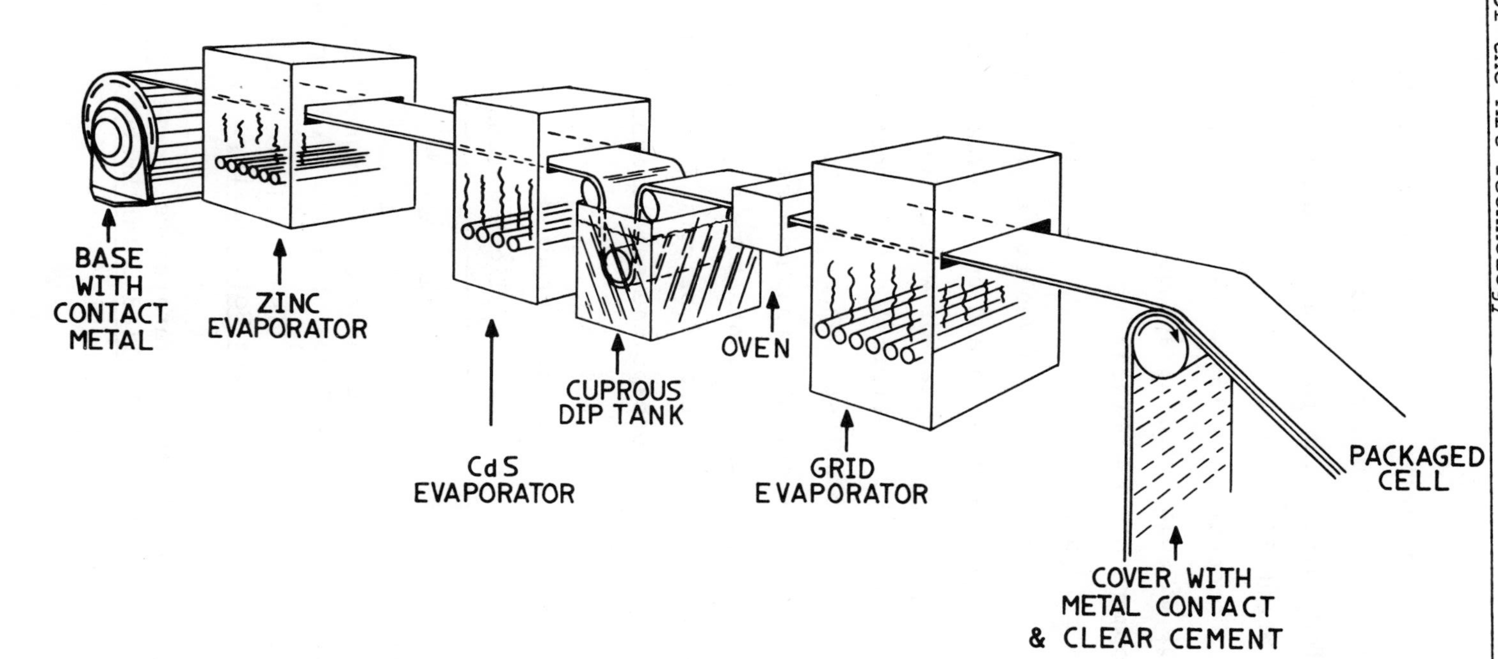

Figure 18. Schematic Representation of a Continuous Process for Manufacturing CdS-Cu_2S Photovoltaic Converters.

sunlight, the material would heat and degrade more rapidly. Such a scheme would necessarily include systems for maintaining the temperature of the CdS-Cu_2S below about 50°C.

Mass production and cost reduction in the fabrication of Si solar cell arrays depends a great deal upon development of new methods of growing large single crystals in a ribbon or thick film. Present techniques do not lend themselves to large returns to scale of mass production and only so much can be done with automation. One factor which may allow considerable reduction of electrical output cost is the ability of Si to withstand heat and to provide increasing current output with increasing intensity of light. Hence, light concentrators may be used which are much less costly than fabricating cells of sufficient area to intercept the same amount of solar energy.

Several researchers have studied the power output of Si cells as a function of light intensity.[49,50,51,52]

---

49. P. A. Berman, _Solar Energy_, _11_, No. 3-4, 180 (1967).

50. M. Wolf and H. Rauschenbach, "Advanced Energy Conversion," Pergamon Press, London, 1963, pp. 455-479.

51. A. Zaitseva and A. Polisan, _Geliotekhnika_, _8_, No. 3, 28 (1972). _Applied Solar Energy_, _8_, No. 3-4, 20 (1973).

52. I. Savchenko and B. Tarnizhevskii, _Geliotekhnika_, _8_, No. 4, 20 (1972). _Applied Solar Energy_, _8_, No. 3-4 83 (1973).

At ambient temperatures as the intensity increases the voltage increases slightly and current flow increases almost linearly for intensities several times that of direct sunlight. Because the internal dissipation of energy across the cell series resistance (bulk, diffused layer, and contacts) increases with the square of the current, cells must be designed to have very low series resistance and good ohmic contacts. With increasing intensity of illumination at constant temperature the efficiency actually increases until the I^2R losses across the cell exceed the gains from voltage enhancement. The optimum is a function of the design of the cell, particularly the junction depth and grid contact.

If cooling is not provided, under increased intensities the solar cells heat and tend to become less efficient. Cells have been reported[49] which under an intensity of approximately four times direct sunlight the temperature reached about 51°C and the efficiency remained constant. Recently, a detailed study of Si photocell output power variation with illumination and temperature was made.[52] The optimum concentration was estimated on the basis of minimum costs per unit power for three cooling methods; radiation, air, and water. The results are summarized in Figure 19, which was derived from reference 52. Curves corresponding to various ratios of output power, p, to the power generated under 80 mW/cm^2 (i.e. 74 W/ft^2) illumination conditions at 0°C, P_0, are plotted in the coordinates of photocell temperature versus illumination. As illustrated, if one is operating under illumination of 2 W/cm^2 (i.e. about 18 times direct sunlight normal to the earth's surface) one could achieve a power output relative to that under 80 mW/cm^2 illumination at 0°C, P/P_0, of 10, 8, or 5 by cooling the cell to 50, 85, or about 135°C, respectively. There are considerable differences in effectiveness of cooling by radiation, air under natural convection, and water. The maximum power increase is a factor of 10-11 for water cooling, 4.5-5 for air cooling, and only about 1.5-1.8 for radiative cooling.

The optimum concentration level, however, is not defined by the maximum output power. Cost factors must be considered. The cost of the system will be the sum of costs of the cells, concentrator, and cooled structure. If the cost per unit concentrator area is 0.2% of the cell cost per unit area, and the structural cost is 1.7% of the cell cost without concentrators,[53,54] the optimum concentration level (least cost per unit output) can be derived[52] as in Figure 19. For water cooling, the cost per unit power can be

53. N. S. Lidorenko, Elektrotekhnika, No. 2, 1 (1967).

54. N. S. Lidorenko, Geliotekhnika, 7, No. 2, 52 (1970).

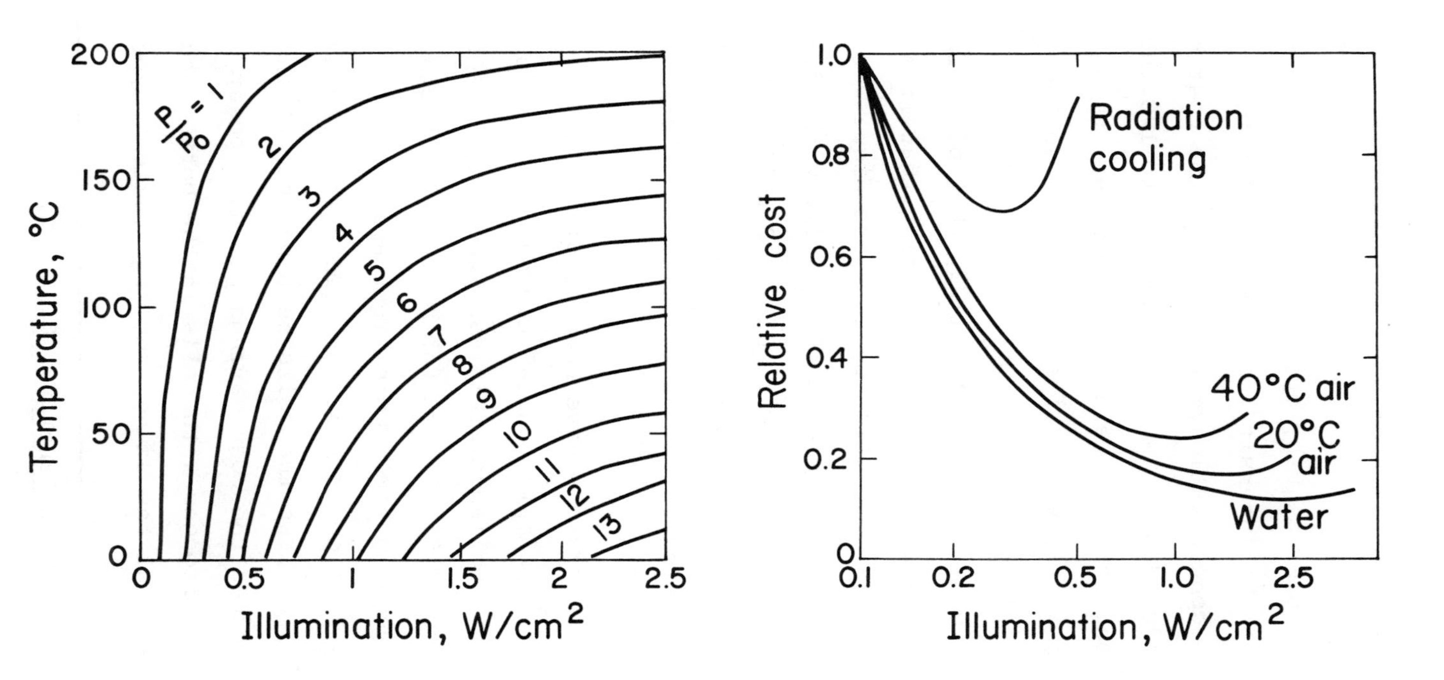

Figure 19. Operating Temperatures and Relative Costs of Silicon Photovoltaic Converters Provided with Sunlight Concentrators and Cooling. Direct Sunlight = 112 mW/cm^2. P/P_o - Output Power/Output Power Under 80 mW/cm^2 Illumination. (Composed with permission of Allerton Press, Inc., Ref. 52.)

reduced by a factor of 8 by operating at 2.5 W/cm^2—about 22 times the intensity of direct sunlight. Hence using a concentration ratio of 22:1 the cost of the solar produced power would be reduced to 1/8 the cost if no concentration were used. However, the maximum power increase is a factor of 10-11 so the concentrator-cell array now operates at an area-effective efficiency of one half, 11/22, that of cells without concentrators. It is interesting that if a factor of 8 in cost reduction by this technique were required to make Si cells cost competitive with CdS-Cu_2S to provide the same power, CdS-Cu_2S cells with 7% efficiency would require the same area as Si cells operating at an area-effective efficiency of 14%/2 or 7%.

Various concentrators have been studied and patented. A very simple conical concentrator can be made with a thin polished stainless steel sheet as the side, with a round Si cell at the base. Such a cone will collect about 3 times more light and without added cooling will operate at the same efficiency as bare cells.[55] Because the cells need not be cut into rectangular shapes the cost savings per unit power output is 80%. Troughs with cells in the bottom and reflective metal at the sides, and "egg carton" type

---

55. E. L. Ralph, Solar Energy, 10, No. 2, 67 (1966).

designs may be used as well as more sophisticated parabolic reflectors.[56,57] A major deficiency in using such devices is the directional nature of their effectiveness. Unless they are oriented toward the sun, the power generated may be much less than for cells not utilizing collectors. A relatively minor problem is that of providing power for the cooling system. Forced circulation water cooling requires on the order of 10% of the power output of the cells.[58]

Storage of Electrical Energy

Surplus electricity generated by photovoltaic arrays during sunlit hours can be stored in common lead-acid batteries for use during nongenerative periods, at least in the small application. However, for large applications such as power plants battery storage appears impractical because of cost and weight. Electrolysis of water to yield hydrogen (H_2) which can be distributed as a gas and burned or utilized in electricity-generating fuel cells offers some advantages. The fuel cells would be operated at the consumer's site. Power can be stored also by pumping water uphill

56. E. Lapin, A. Ernest, and P. Sollow, U.S. Patent 3,427,200; February 11, 1969.

57. N. Regnier and M. Shaffer, U.S. Patent 2,919,298; December 29, 1959.

58. W. Beckman, P. Schoffer, W. Hartman, Jr., and G. Lof, *Solar Energy*, *10*, No. 3, 133 (1966).

and allowing its downward flow to run a turbine (pumped water storage), or by turning a flywheel, or various other methods.

Of most convenience currently is the lead-acid battery, at least for small applications. The cost per kiloWatt is about $100 and a battery can be expected to last 5-7 years under recommended recharging rates.[3] To illustrate this, a Sears DieHard 27C battery is retail priced at $32.95 with trade in.[59] It has a reserve capacity to be discharged at 25 amps for 170 minutes and maintain a voltage of 10.5-12 volts. If the average voltage is 11.2 V the battery delivers an average of 280 W for a period of 2.83 hrs. Then the price per kiloWatt is $116 at low volume retail prices. Applying this to the storage of electricity required for a solar cell powered residence; if the average electricity usage[60] is 8000 kWhr/yr the average requirement for a two day period would be 44 kWhr. One could expect (44000/280)/2.83 or 55 fully charged batteries to supply the electricity at 15.4 kW maximum capacity, i.e., 140 amps in a 110 V circuit assuming 100% transformation efficiency. The price of the batteries would be about $1812. If the cost of money to the average householder is 10% and the

---

59. Sears Fall and Winter Catalog, 1973, p. 779 (Technical Data).

60. See reference 6, Chapter 2.

batteries last 6 years, the cost of average two-day storage capacity is $535 per year. At the 8000 kWhr/yr usage rate, this amounts to 6.7 cents per kWhr—about three times the price of utility-provided electricity in 1972.

Fuel cells, in particular the hydrogen-oxygen cell, offer the advantage of higher energy output per weight than batteries.[61] They are devices which generate electricity by constraining the oxidation of a fuel, H_2 in this case, in such a way that electricity must flow through an external circuit for the reaction to occur.[62] The principles can be schematically shown as follows for two types of cells.

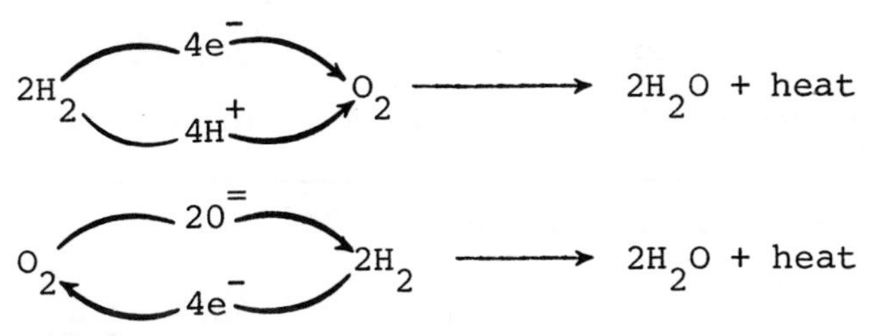

Although fuel cells offer the advantages of compactness and high energy density, raw materials must be provided safely. If portability is required, there is some disadvantage to moving a bulky tank of H_2.

61. K. R. Williams, "An Introduction to Fuel Cells," Elsevier Publishing Company, New York, 1966, p. 134.

62. S. W. Angrist, "Direct Energy Conversion," Allyn and Bacon, Inc., Boston, 1965, p. 326.

The efficiency of storage of photoelectric power by electrolysis of water to produce H_2 and O_2 (90-95%)[63] and then utilizing the H_2 and O_2 to generate electricity in a fuel cell (60-80%)[61, 63] is about 65%, while lead-acid batteries are about 70% efficient.[64] However, fuel cell power is about twice as expensive[64] and needs development to come into general use.[65] Fuel cells suffer somewhat from a short life, relatively high cost, and scarcity of catalytic metals.[66,67] With sufficient foreseeable development it is believed that electrolysis, H_2 storage, fuel cell regeneration of electricity, and inversion, will be a practical method of storing energy for use in peak period generation of commercial electricity.[68] The H_2 would be stored in natural or man made caverns

---

63. See reference 17, Chapter 1.

64. F. Daniels, "Energy Storage Problems," Solar Energy, 6, No. 3, 78 (1962).

65. W. Grubb and L. Niedrach, "Fuel Cells," in Direct Energy Conversion, Edited By G. W. Sutton, McGraw-Hill Co., New York, 1966, p. 39-104.

66. R. Boll and R. Bhada, Energy Conversion, 8, No. 1, 3 (1968).

67. B. Baker, Fuel Cell Systems-II, "Advances in Chemistry Series, Edited by R. F. Gould, American Chemical Society Publications, Washington, D. C., 1969.

68. A. Bruckner, II, W. Fabrycky, and J. Shamblin, IEEE Spectrum, April, 101 (1968).

and utilized upon demand. Thus, the main generating plants could be designed to operate with less reserve capacity, at a higher load factor, and at lower capital cost. If this proves practical for use with conventional plants it probably would serve well in conjunction with a solar energy plant.

Although it appears that storage of photo-generated electricity would be very expensive relative to what is presently paid for on-demand electricity from a utility, this cost is minor in many applications such as those in remote areas or those requiring portability. Furthermore, the cost of electrical storage can be expected to decrease in the future with more research and development by existing institutions aimed at new markets which require lower cost, more energy-dense storage.

Chapter 5

ECONOMIC CONSIDERATIONS IN THE DEVELOPMENT OF PHOTOVOLTAIC ENERGY CONVERSION

Because photovoltaic energy conversion technology is essentially in the research and prototype development stage, it is difficult to predict accurately what the cost of mass produced converters for terrestrial use would be. It is even more difficult to predict the rate of discovery and technological progress which would allow reduction of costs. Much depends upon the incentives for exploration of economical processes for mass producing long-lived converters with increased efficiency. The present state-of-the-art was achieved primarily by the personal interest and drive of a relatively small group of researchers in individual efforts. Some incentive for production of high-efficiency stable silicon cells was provided by the space efforts. However, the alternative sources of energy are minimal and costly in such remote locations, large manufacturing capacity which would yield increased returns to scale has not been needed, and the profits have not provided incentive for large efforts by private industry. Direct governmental funding of research in solar energy started in the 1970s in a very modest fashion[1,2] but may increase to a more effective amount as the energy

1. J. Denton and L. Herwig, Proceedings of the 25th Power Sources Symposium, May 1972, p. 137.

shortage becomes more critical. As so aptly stated by M. L. Kastens,[3] "...the world of science, and certainly the world of industry, looks somewhat askance at the enthusiastic proponent of research in solar devices."

Of course the askance look is brought about by the economics of commercialization of photovoltaic devices. Attention to this field by people skilled in the techniques of design and production could reduce costs. Mass production techniques could introduce economies of scale at the production level. But mass production requires mass markets and mass markets are not easily created in fields such as this where so many other inexpensive and convenient sources of energy have been developed. Various psychological deterrents are also at work. Who wants to have electricity at the whim of the weather? Who wants to have a lot of black solar collectors on his home? Who wants to, or has the money to make that large capital investment now in return for "free" energy in the future? The psychology of the marketplace certainly will influence the future commercialization of photovoltaic energy converters.

---

2. National Sciences Foundation Legislation, Hearing before the Special Subcommittee on the National Science Foundation, May 3, 1973, U.S. Printing Office, Washington, D. C., 1973.

3. M. L. Kastens, "The Economics of Solar Energy," Introduction to the Utilization of Solar Energy, Edited by A. Zarem and D. Erway, McGraw-Hill Book Co., New York, 1963, pp. 211-238.

If conventional sources of energy become unreliable relative to individually-owned energy supplying devices, people will pay an appropriate price for the relative security of their own energy supply.

Let us consider the inherent characteristics of solar energy which define many of the economic parameters of photoelectricity generation. During photoelectric generation, no liquid or gaseous wastes are generated, no unusual safety problems arise, no radioactivity is released, and for most uses, no thermal problems are encountered other than normally associated with a dark roof. Space used for solar collection may be used for other purposes as well, such as roofs, walls, and surfaces of reservoirs. The only obvious associated environmental problems are the possible visible undesirability of dark surfaces and any problems of safety associated with battery, fuel cell, or other means of storage of energy.

Because the "fuel" for a solar energy converter is free, the cost of power output is nearly independent of the conversion efficiency as long as plenty of cost-free area for collectors is available. The most significant economic factor is not output per area, but output per unit of capital cost properly adjusted for the time value of capital, depreciation, taxes, and the relatively minor maintenance costs. Unless space is limited it is a minor matter that one material is 7% and the other is 14% efficient if the required investment in the former is the same as for the latter to achieve a given supply of power.

Because capital depreciation continues at the same rate regardless of the usage of the solar converter output, the load factor (percent of time full capacity is used) is very significant. Solar energy on the ground is a double periodic function with periods of 24 hours and 365 days onto which is imposed the random fluctuations of cloud cover. The maximum load factor for photoelectricity generators would be about 40% in places like Phoenix and Las Vegas (see Figure 6). The average load factor over the U.S. would be 16%, i.e., 17 $W/ft^2 \div 104\ W/ft^2$; the average solar incidence divided by the incidence from direct sunlight on a plane normal to the sun. Because the expense does not increase appreciably when the sun is not shining, the cost per unit output varies in a nearly linear inverse manner with the load factor. If all other factors such as degradation rate and maintenance remain constant, terrestrially generated photoelectricity in Phoenix would cost about 40%, i.e., 16/40, of the average cost across the U.S.

Certainly the cost of photovoltaic arrays should decrease with mass production. However, unless there is a quantity discount, the small consumer would not realize economies of scale in building larger arrays. The output of a collector is directly proportioned to its area. Hence, more capacity would be accomplished by simply adding more modules of the same photovoltaic materials. Indeed, structural complications may lead to decreasing returns to scale. This is in sharp contrast to most other forms of electricity generation. Thermoelectric and hydroelectric generating stations provide less expensive power as the size increases. Diesel generated electricity used in remote locations, locomotives, etc., is much less expensive the higher the capacity. This is possible by the capital cost savings as unit size increases. Because small scale photoelectricity generation is as cost effective as large scale, the cost of distribution of power from central locations can be avoided with photovoltaic energy conversion at the site of electricity utilization. The diffuse nature of solar energy lends itself to utilization in diffuse and remote areas.

The attributes of solar energy and photovoltaic energy conversion must be considered in relationship to those of conventional public sources of electricity which require massive investment to realize the returns to scale, visual undesirability of power lines, utilization of fossil fuels which could go into petrochemicals, fertilizers, and plastics, and air, water, thermal, and in some cases radioactive pollution. These are predominantly societal expenses however, which cannot be fully assessed by the individual consumer when he is deciding whether to connect into a commercial power system, which provides a reasonably secure supply of power at about $180 per year, or to construct his own supply at an immediate cost of several thousand dollars and future responsibility for its maintenance and operation.[4] The two most fundamental considerations by the consumer are the expected costs of commercial power (which will vary with fuel costs) and the initial investment in solar conversion equipment.

It is a reasonable deduction that photovoltaic energy conversion will be utilized only as it becomes cost competitive with other sources of electricity in given market segments. It is cost competitive with primary (non-rechargeable) batteries and fuel cells for long unattended usage in space. It is also cost competitive in remote locations such as telephone repeater stations, lighthouses, and buoys where delivery of fuel or replacement of primary batteries is difficult. It will be only through much lower costs of photocells that the mass markets for electricity can be penetrated.

The novelty and intrigue of such a direct clean method of electricity production may lead to sincere but optimistic conclusions concerning the viability of the business of manufacturing photovoltaic devices. The visionary of today may be the disillusioned dreamer of tomorrow if the marketplace is not realistically

4. P. Glaser, Testimony before the House of Representatives, reported by J. Thomas, Boston Globe, January 20, 1974, p. A-3.

evaluated in light of business objectives and technological possibilities. The following discussions address some of the related factors. Of course demand and supply of electricity is crucial to future development of photovoltaic energy conversion devices.

Demand for Electricity, Supply, and Costs

Conventional - Nearly all of electricity used in the U.S. is supplied by large electric utilities. Exceptions to this occur in cases where portability or remote usage is required. Primary batteries (those used once and discarded) such as the carbon-zinc LeClanche "flashlight," alkaline-manganese, and mercury cells are used for powering flashlights, toys, camera photoelectric eyes, transistor radios, clocks,...etc. Secondary batteries (rechargeable) such as lead-acid and nickel-cadmium are used for starting, lighting, and ignition service in automobiles, boats, and aircraft, and to power electric toothbrushes, shavers, fork lift trucks, golf carts,...etc. Secondary batteries are generally recharged by electric power from electric utility companies or from engine powered generators. Still other electricity is generated by diesel or gasoline engine driven generators, particularly of the industrial size used at remote construction sites.

The present annual usage of utility-provided

electricity is about 2 x 10^{12} kWhr.[5,6] With internal usage and sales of about 1.8 x 10^{12} kWhr at an average price of about 1.8 cents per kWhr, this is a 32 billion dollar business. It is expected to grow at an average annual rate of about 6.9% for the next 10 years and probably until the end of the century.[7] The growth in electricity usage is due in large part to the cleanliness and convenience of its utilization. Convenience, reliability, and cleanliness are certainly of prime importance to a customer. Electricity offers energy-on-demand by the flip of a switch. Its growth has been enhanced also by a general decrease in prices[5] until 1970 as illustrated in Figure 20, and increasing electrification of homes in remote areas.[8] Although nearly 100% of the U.S. homes are now supplied with electricity, large percentages (over 50% in 1970) do not have air conditioners, electric water heaters, ranges, dryers, dishwashers, freezers, disposals, blenders, and electric space heating. The

---

5. See reference 6, Chapter 2.

6. ITC Report C645, "The U.S. Energy Problem," Volume I Summary Volume, Intertechnology Corporation, Warrenton, Virginia, 1971.

7. See reference 4, Chapter 1.

8. O. L. Culberson, "The Consumption of Electricity in the United States," Oak Ridge National Laboratory, 1971, ORNL-NSF-EP-5.

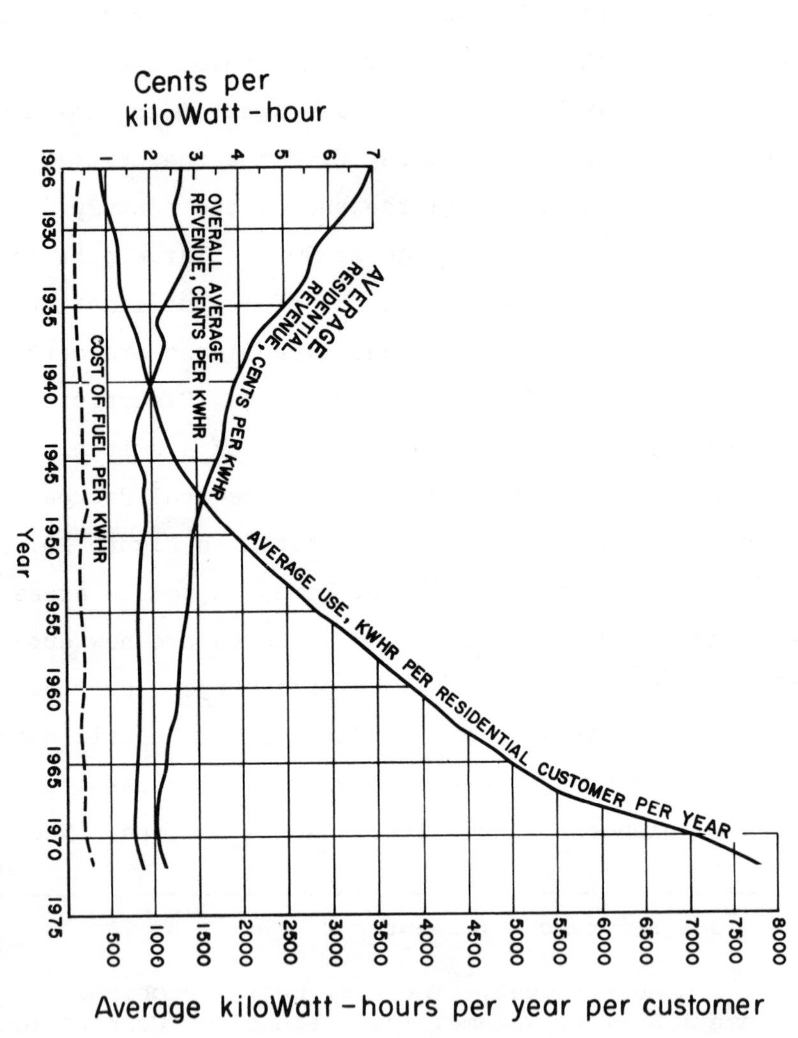

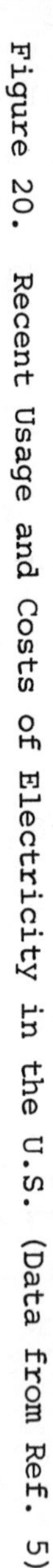
Figure 20. Recent Usage and Costs of Electricity in the U.S. (Data from Ref. 5)

residential usage per customer has increased at an average annual rate of 6.5% over the past 20 years (see Figure 20). About one third of U.S. electricity is used in residences, about 22% in the commercial buildings, and nearly all the rest in industry.[7,8,9] Aggregating by the conventional sectors; residential and commercial 55%, industrial 41%, and others 4%.

Because of various efficiencies associated with predictable high quantity usage, the price of electricity paid by the industrial sector is about one half that by residential and commercial customers. In 1970 the average prices were 2.10, 2.01, and 0.95 cents per kWhr for residential, commercial, and industrial customers, respectively.[10] The 1970 National Power Survey estimated that the average cost of electricity in constant dollars would increase by 19% between 1968 and 1990.[11] Even though this seems to be a low estimate in light of recent oil price increases, there are valid reasons to predict that electricity prices will not increase exorbitantly. (I will later assume a quadrupling of fuel costs for my own upper estimate.)

---

9. J. Tansil, "Residential Consumption of Electricity 1950-1970," Oak Ridge National Laboratory, 1973, ORNL-NSF-EP-51.

10. Federal Power Commission, "1970 National Power Survey," Washington, D. C., Volume IV, 1971, p. 3.

11. *Ibid.*, Volume I, p. 19.

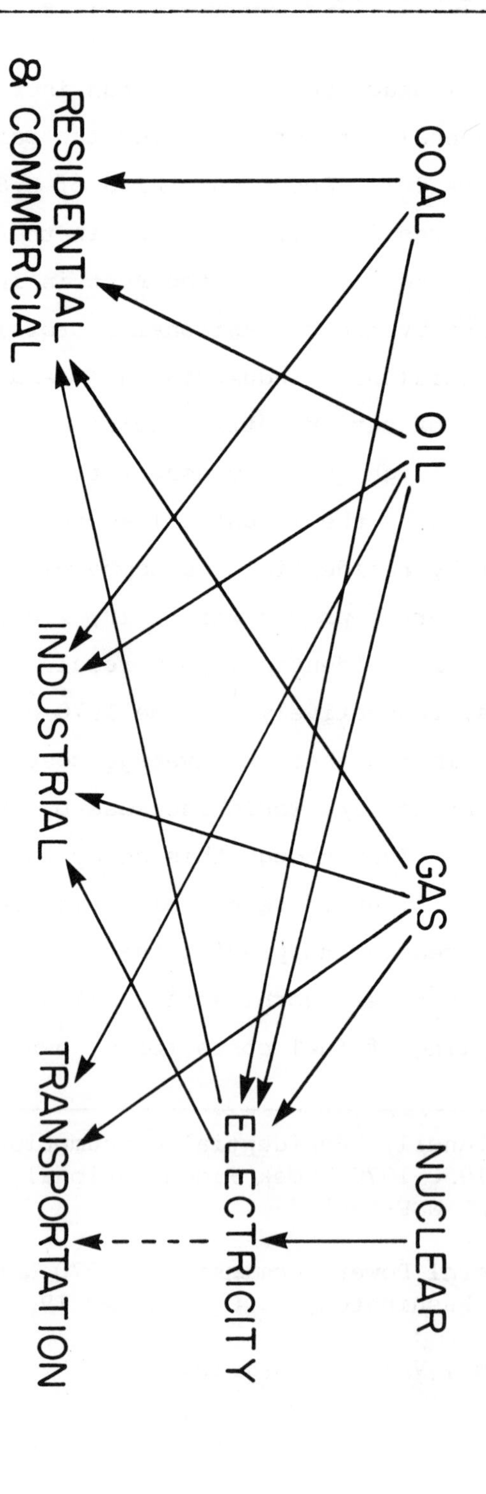

Figure 21. Interfuel Substitutability for Electricity Generation.

As illustrated in Figure 21, the energy flows in the U.S. energy system afford use of all the primary fossil or nuclear fuels in electricity generation.[12] This allows considerable interfuel substitutability in the electric utility industry. In 1970 electricity generating plants were powered by coal 56%, oil 14%, gas 30%, and a small percentage by nuclear fuel. Nuclear generation has and will increase the interfuel substitutability.[13,14] About one half of the power stations have multiple fuel capability. In 1969, 9.9% could burn coal and oil, 10.4% coal and gas, 19.2% oil and gas, and 8.3% coal, oil, and gas. Those using only one fuel were coal 40.8%, oil 2.3%, gas 7.2%, and nuclear 1.8%. Nearly 70% of the generating capacity could be powered by coal. In recent years air pollution regulations have resulted in even greater flexibility because many coal burning plants have been

---

12. D. C. White, Energy Laboratory, "Final Report Submitted to the National Science Foundation—Dynamics of Energy Systems," M.I.T., Cambridge, Massachusetts, 1973.

13. T. D. Duchesneau, Federal Trade Commission Economic Report, "Interfuel Substitutability in the Electric Utility Sector of the U.S. Economy," U.S. Government Printing Office, Washington, D. C., 1972.

14. A report of National Economic Research Associates to the Edison Electric Institute, "Fuels for the Electric Utility Industry 1971-1985," Edison Electric Institute, New York, 1972, EEI Pub. No. 72-27.

converted to use gas or residual fuel oil and have maintained the coal facilities.

An interfuel substitutability study[13] has shown that the fuels used in various sections of the nation are a function of their price and availability. Because primary fuel prices increase at varying rates, the changing relative prices among fuels can be expected to alter the mix of fuels consumed.[15] Most regions have two competitive primary fuels and some have more. Of course not all regions have identical supply factors nor labor and capital costs so the price of electricity varies from place to place as illustrated in Table 2.[16] The introduction of nuclear generation should tend to reduce the higher values relative to others.

The conventional electricity generation system may be schematically represented in a systems dynamics[17,18] model as shown in Figure 22. The sign beside each

---

15. J. Griffin, Bell J. Econ. Mgmt. Sci., 5, 515 (1974).

16. Data derived from: Edison Electric Institute Statistical Yearbook of the Electric Utility Industry for 1972, New York, No. 40, Publication 73-13.

17. J. W. Forrester, "Principles of Systems," Wright-Allen Press, Cambridge, Ma., 1972, Second Preliminary Edition.

18. J. W. Forrester, "Industrial Dynamics," M.I.T. Press, Cambridge, Ma., 1961.

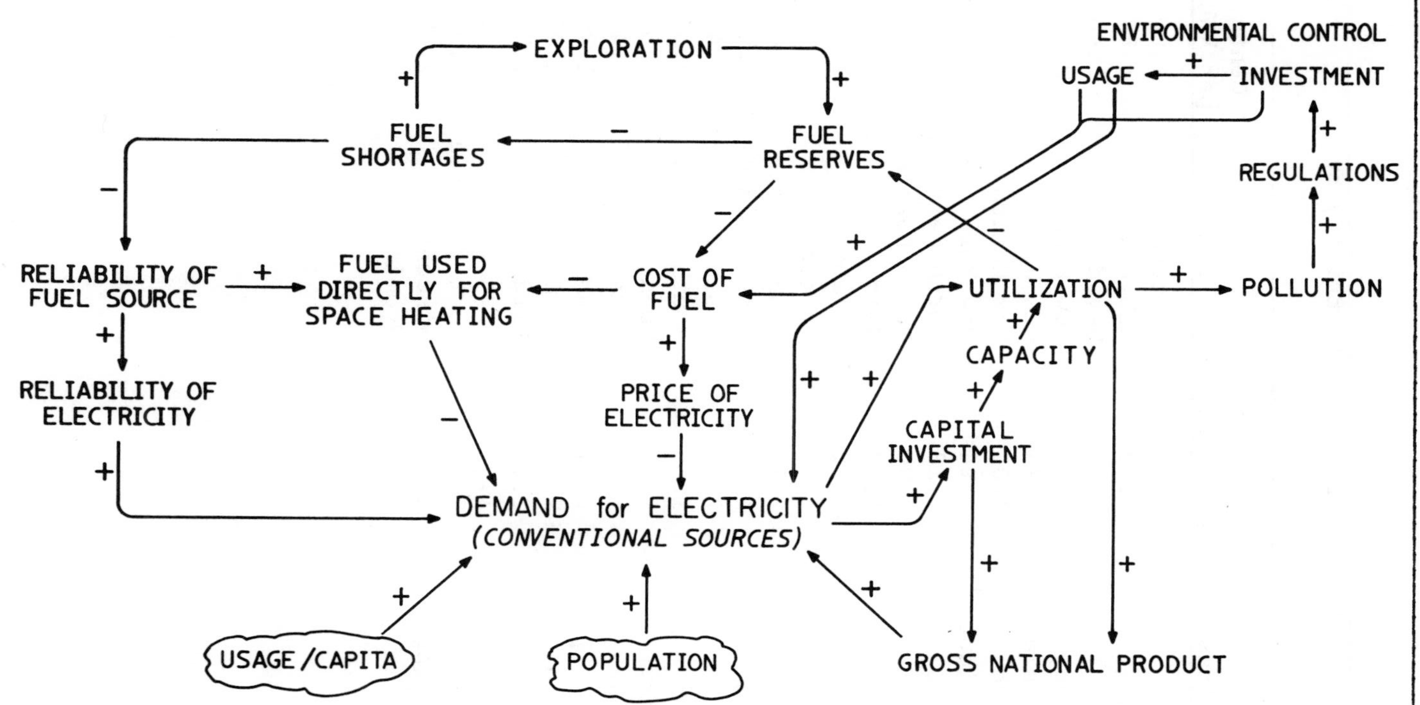

Figure 22. Schematic Representation of the Conventional Electricity Generation System. An Increase in the Level of a Cause (Tail of Arrow) Results in an Increase (+) or Decrease (-) in the Level of an Effect (Point of Arrow).

Table 2. Average Price of Electricity by States in 1972

States	Cents per kWhr	States	Cents per kWhr
Alabama	1.26	Montana	1.00
Alaska	3.20	Nebraska	1.84
Arizona	1.88	Nevada	1.40
Arkansas	1.69	New Hampshire	2.22
California	1.73	New Jersey	2.46
Colorado	2.02	New Mexico	1.96
Connecticut	2.44	New York	2.58
Delaware	2.13	North Carolina	1.54
Florida	1.96	North Dakota	2.37
Georgia	1.67	Ohio	1.59
Hawaii	2.38	Oklahoma	1.80
Idaho	1.07	Oregon	0.98
Illinois	2.11	Pennsylvania	2.04
Indiana	1.78	Rhode Island	2.56
Iowa	2.21	South Carolina	1.41
Kansas	1.85	South Dakota	2.39
Kentucky	1.15	Tennessee	1.03
Louisiana	1.47	Texas	1.47
Maine	2.17	Utah	1.79
Maryland	2.12	Vermont	2.23
Massachusetts	2.66	Virginia	1.71
Michigan	1.93	Washington	0.72
Minnesota	2.10	West Virginia	1.54
Mississippi	1.65	Wisconsin	2.08
Missouri	2.11	Wyoming	1.60

arrow represents a decreasing (-) or an increasing (+) effect of the causing parameter at the tail of the arrow on the affected parameter at the point of the arrow. The demand for electricity from conventional fuels is governed by economic, population, usage rate, pollution, price, and reliability parameters. Such a model could be drawn for each primary fuel source with integration of models to reflect the dynamics of cross-over from one fuel to another as respective supplies and prices changed. Cross-overs such as the following would affect the price, reliability, and pollution associated with electricity generation to meet consumer demand.[14]

1. Stack-gas desulfurization vs. low-sulfur coal and oil prices.
2. Uranium vs. fossil-fuel prices.
3. Gas vs. low-sulfur coal and oil prices.
4. Purchase low-sulfur oil vs. topping of high-sulfur oil.
5. Shale oil vs. domestic and inflated foreign oil prices.
6. Synthetic gas from coal vs. natural gas.
7. Distillate oil vs. gas supply and price.

As the price of one fuel increases the demand for others would increase at the expense of the former. Increased reliability of supply and decreased pollution associated with a particular fuel would increase the demand for it relative to the others.

In the next 10-20 years, stack-gas desulfurization is expected to be technically feasible in an economic[7,14] way, adding about 0.1 cent/kWhr to the price of electricity. Uranium powered generation is expected to be less expensive than fossil fueled generation, hence keeping electricity prices down. Gas is expected to be more expensive and less available for generating purposes. Refining of low quality crude oil by utilities will make more suitable oil and synthetic gas. Under expected policy conditions oil shale will provide a long term ceiling of about $8-12/Bbl for rising real costs of imported and domestic oil. Distillate oil for residential heating is expected to increase in price more slowly than gas. And synthetic gas production from coal will hold the price of gas below about $1 per thousand cubic feet (3 times the 1974 average price) while increasing the price pressure on coal.[14]

If the above generally accepted predictions prove to be correct, it would seem improbable that pollution abatement requirements and shortages of particular fuels could drive the constant dollar cost of fuel plus capital costs associated with pollution abatement to as much as four times their 1970 values by the end of the century. Short term adjustments to meet demand may make this estimate seem low during periods when demand is exceeding supply due to time lags in the system. However, an increase of a factor of four in cost elements is a very high long term estimate by comparison to recent projections,[7,11,14,19] and should be considered applicable only under extremely unfavorable economic conditions for development of energy sources.

The average price of electricity in 1970 was 1.59 cents/kWhr of which 0.33 cents was for fuel.[5] If other costs remain constant, pollution abatement capital and operating costs add 0.2 cents/kWhr, and fuel costs quadruple, the average cost of electricity should not exceed 2.8 cents per kWhr in 1970 dollar values. If the relative prices in the consuming sectors remain as in 1970 by the year 2000 the average price for residential electricity should not exceed 3.7 cents/kWhr, commercial 3.4 cents/kWhr, and industrial 1.7 cents/kWhr. Hence, large increases in electricity prices from conventional supplies are unlikely and will not significantly impact the probability for development of solar converters.

Photovoltaic - The demand for electricity is a derived demand. Consumers buy it to satisfy other demands for light, heat, refrigeration,...etc. Energy in the form of electricity is rarely visible to the consumer. Therefore, consumers will make decisions about electricity source predominantly on the basis of cost, availability, convenience, reliability, and the timing

---

19. See reference 16, Chapter 1.

of monetary outlays. To compete favorably with fossil-fuel generation for the largest markets of electricity usage photovoltaic generation must offer electricity at a comparable price, with provision for continuous availability and equal reliability. With proper financial and ownership arrangements the monetary flows could be arranged to follow patterns suitable to the consumer, although if the consumer owns his own photovoltaic facility the initial outlay, or the realization of what the long term outlay is, may present a formidable barrier to marketing of photovoltaic converters—even if the produced electricity is as inexpensive as buying electricity from conventional utilities.

A diagram of the factors which govern the development of photovoltaic electricity production and the demand for it is presented in Figure 23. For photoelectricity to compete favorably for the entire market, the price and reliability must be comparable to electricity from conventional sources unless pollution which is recognized as being generated along with electricity becomes a major detriment in consumer buying or electricity availability. The price must reflect the cost of photovoltaic array manufacturing and selling, profit, installation and cost of capital, as well as the cost of maintenance and electrical storage to make the supply reliable. At 10% conversion efficiency a horizontal one square foot photovoltaic device would

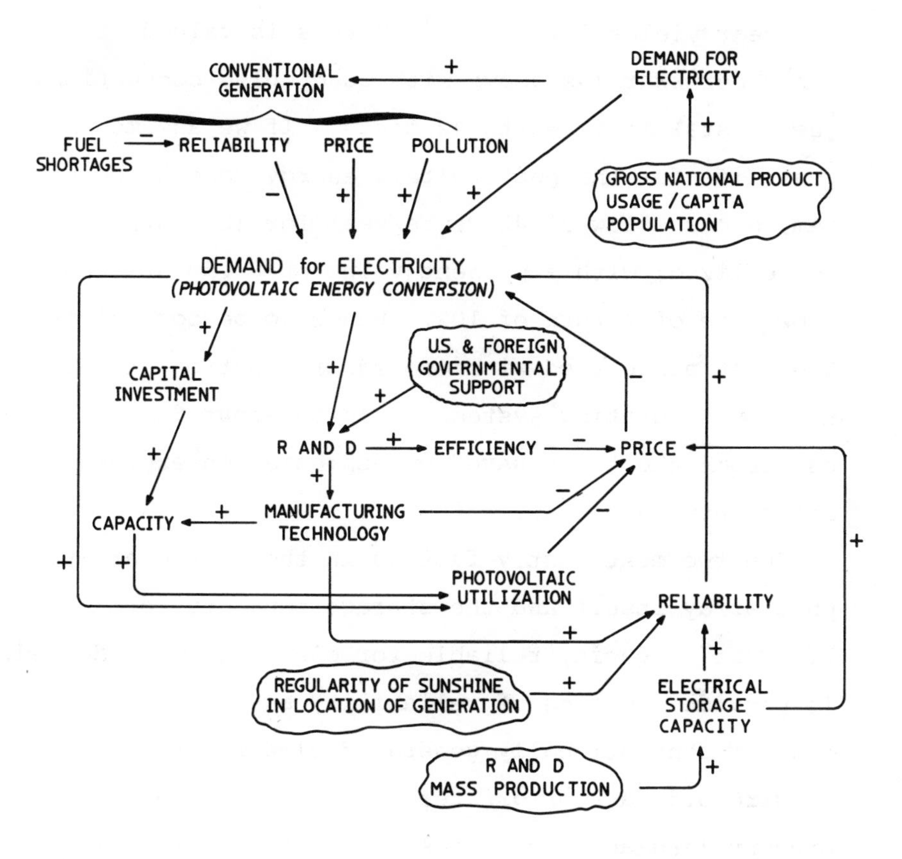

Figure 23. Schematic Representation of the Factors which Affect the Development of Photovoltaic Converters. See Explanation of Notation in Figure 22.

produce an average of 1.7 W of electricity averaged annually over the U.S. (this may be higher with different orientation). This multiplied by 8766 hours per year yields 14.9 kWhr. If this is valued at 2.8 cents per kWhr (as above with quadrupled conventional fuel costs) it is worth 42 cents. If we assume a 10 year life for the photovoltaic energy generation system, a cash flow of 42 cents/year for 10 years would be realized, with a present value of $2.58 at a cash flow rate of return of 10%. Hence to be competitive the cost associated with one square foot of the electricity generating system should be about $2.58. Let us assume the taxes saved by depreciation expense offset maintenance costs.

The two most costly factors in the system are the photoarray itself and the storage capacity to make 1.7 W/ft^2 capacity reliable for electricity-on-demand. To provide a degree of reliability anywhere close to that of conventionally generated electricity would require 3.5 days or greater storage capacity. With present lead-acid batteries the cost[20,21] associated with each square foot of array would be about five dollars. Fuel cells would cost at least double that.[22]

20. See reference 2, Chapter 2.

21. See reference 59, Chapter 4.

22. See reference 64, Chapter 4.

Present solar arrays used in space[20] cost \$1000-\$8000/ft^2. Some photocells are available[4] for \$20 per Watt of continuous power: W_c, (100% load factor). For the 10% efficient one square foot array under consideration which would operate at a load factor of 16%, i.e., 17 W/ft^2 average light incidence divided by the maximum ground level intensity 104 W/ft^2, one would have to pay \$20/W_c x 1/0.16 x 1.7 W/ft^2 or \$208 per ft^2. Hence photovoltaic energy conversion is not economically viable in the large electric utility business at today's prices. The cost of photocells would have to be reduced by a factor of 100-500, and the cost of electrical storage by a factor of about 5-10 before it could be cost competitive.[22]

Of course present manufacturing capacity is many orders of magnitude below what would be required if photovoltaic converters were viable in this market. Production of the present annual usage of 2 x 10^{12} kWhr would require 230,000 megaWatts. With allowance for reserve capacity and storage efficiency, a 500,000 megaWatt capacity would be required. At 10% efficiency about 300 billion square feet of photovoltaic material would be needed. If the replacement rate were once every ten years, a manufacturing capability of 30 billion square feet per year would be required.

It appears that considerable research, perhaps stimulated by governmental funding will be required to make photovoltaic electricity competitive with conventionally generated power. From the above cost considerations one can derive that terrestrial photovoltaic electricity generated with today's technology would cost about $2.30/kWhr as compared to the average utility price of $0.018/kWhr.

Markets for Photovoltaic Energy Conversion

From the above cost and supply considerations it is obvious that photovoltaic electricity generation is economically infeasible for large scale use, and probably will be for the rest of the century unless very great advancements are made in both solar cell and energy storage technologies. However, this does not eliminate photovoltaic usage in smaller segments of the overall market.

Several markets involving use of batteries and small internal combustion engine driven generators in remote locations can be considered. Although small in relation to the entire electricity market, they may provide some incentive for commercialization of specialized solar power packages. Photocells might also be used to supplement conventional supplies of electricity during peak periods of usage if their cost can be reduced, and no storage is required. Photocell technology which is in its infancy is probably more apt to achieve major cost reductions than is the relatively mature electrical storage industry.[23] Hence applications not

23. C. L. Mantell, "Batteries and Energy Systems," McGraw-Hill Book Co., New York, 1970.

requiring storage may be of particular interest as the price comes down. We might separate markets into three sizes; megaWatt, kiloWatt, and Watt.[24]

MegaWatt - As illustrated above, both solar cell and electrical storage technology would have to progress exorbitantly to make terrestrial photovoltaic power generation competitive with conventional means at megaWatt levels. Conceptions for space,[25,26] and balloon[27] stations (above many of the light screening functions) with microwave transmission of electrical energy back to ground level appear out of the question as a practical and political matter within this century.[28] One large market may be open to photoelectricity, however, if the price without storage

---

24. H. Tabor, Solar Energy, 6, No. 3, 89 (1962).

25. J. Mockovciak, Jr., "A Systems Engineering Overview of the Satellite Power Station," Seventh Intersociety Energy Conversion Engineering Conference, Conference Proceedings, September 1972, p. 712.

26. P. E. Glaser, "The Future of Power from the Sun," Intersociety Energy Conversion Engineering Conference 1968 Record, August 1968, p. 98.

27. W. R. Cherry, "A Concept for Generating Commercial Electrical Power from Sunlight," Conference Record of the 8th IEEE Photovoltaic Specialists Conference, Seattle, August 1970, pp. 331-337.

28. N. R. Sheridan, Solar Energy, 13, 425 (1972).

can be lowered to within the realm of conventionally supplied electricity, i.e., solar array cost on the order of $2-3/ft^2. This is the production of electricity for peak power usage.[29] Typically about 30-40% more electricity is required in the months of July and August[5] than in other months because of air conditioning. Also more electricity is consumed from about 10:00 A.M. to 7:00 P.M. than any other part of the day. Utilities have to provide excess generation capability to meet peak power requirements and usually operate below 65% peak load factor.[5] Often peak load gas turbine generators and diesel powered generators are used in conjunction with the conventional steam driven generators to meet these demands. The capital investment in conventional steam driven generators is about $120-170 per kW of power. Peak load equipment ranges from $70-250 per kW with an average of about $200 per kW.[7,28] The annualized cost of capital and maintenance account for about one half of the cost of generation of electricity so a solar unit which operates with free fuel could compete at a capital investment of about $400 per kW at peak power (full sunshine). At 10% efficiency this means an operational array could cost a total of about $400/kW x 10.4 W/ft^2 or four dollars per square foot.

---

29. E. L. Ralph, "A Plan to Utilize Solar Energy as an Electric Power Source," Solar Energy, 13, 326 (1972).

Fortunately the peak power demands occur during times of maximum sunshine for the most part. Indeed the demand for electricity for air conditioning is derived considerably from the incidence of sunlight. Peak power photovoltaic systems would necessarily be distributed regionally in a power system grid to minimize the fluctuations from abrupt localized cloud cover. Perhaps the ideal would be to have them mounted on residential and commercial building roofs and feed their users directly when demanded while supplying power to the system when not used at that location. Expenses of disperse inversion and interconnection equipment would be offset somewhat by savings of distribution costs from central facilities which generally amount to about 30% of the cost of capital investment.

Because power from the photovoltaic array would only be needed in approximate proportion to the solar energy incident, storage costs would be minimal. At present, the cost of this peak power with direct sunlight would be $20/W.[30] Because it is only the peak usage periods under consideration, which coincide with peak power output of solar cells for the most part, $20/W is the appropriate cost factor. Hence, the cost of photovoltaic arrays would have to be reduced by a

---

30. Dr. Elliot Berman, President, Solar Power Corporation, Braintree, Massachusetts, Private Communication, February 1974.

factor of \$20,000/kW ÷ \$400/kW, or 50 times, to be competitive for this application. If 10% of the present 400 billion Watt capacity were converted to peak power solar cells with a ten year life, the replacement market would be about 400 million ft^2/year.

At this time absorption of solar energy to form heat to run a steam powered generator is more economically feasible for this application and may remain that way in the future. Some estimates have been made[31,32] that electricity at a cost of about 10 cents/kWhr can be generated this way in ideal locations. However, cooling problems may affect this concept deleteriously,[33] and the cost estimates are not settled.

KiloWatt - The idea of supplying photoelectricity to a household and integrating this generation into the commercial supply system has been subject to detailed

31. G.O.G. Lof, "Engineering and Economic Problems in the Production of Electric Power from Solar Energy," World Power Conference on Energy, Rio de Janiero, 1954.

32. A. B. Meinel and M. P. Meinel, Bull. At. Scientists, 27, No. 8, 32 (1971). Also: "Solar Energy for the Terrestrial Generation of Electricity," Hearings before the Subcommittee on Energy, June 5, 1973, U.S. Printing Office, Washington, D. C.

33. L. J. Haworth, "Assessment of Energy Technologies Report on Step 1," Associated Universities, Inc., Upton, N.Y., 1971, p. 103.

analyses.[30,34,35] A study[34] of the climatic and technological parameters incident to providing continuous electricity for residential use in Phoenix showed that a typical residence readily could be supplied by photovoltaic conversion of sunlight falling on its roof with a 19.5 kW fuel cell storage system to provide storage. Combination of several residential systems would tend to decrease costs and improve reliability because of averaging of peak power usages. Although the technology required in solar cells, fuel cells, and inversion to 110 V ac is available for this purpose it is much too expensive to compete with conventional methods of supplying electricity. Homes or cabins in wilderness areas where alternative means of electricity supply are limited will probably provide the first practical markets for independent residential photovoltaic electricity generation systems.

The cost of residential photovoltaic conversion systems would have to yield electricity at a competitive price for wide usage, i.e., about 2-4 cents/kWhr

---

34. C. E. Backus, "A Solar-Electric Residential Power System," Seventh Intersociety Energy Conversion Engineering Conference, Conference Proceedings, September 1972, pp. 704-711.

35. K. W. Boer, Boston Globe, January 20, 1974, PA-2, Reported by J. Thomas.

in 1970 dollars. As seen above at 10% efficiency, this would mean the total price per square foot of installed operational array would need to be 2-4 dollars/ft^2. This would appear to be a very difficult goal to achieve within this century if storage capacity of up to four days (even in Phoenix)[34] is required. However, if such generation could be made supplemental to normally available utility power and if the residential photovoltaic system could be integrated into the utility grid so no storage would be required,[36] much of the peak power generation for usage in the residential sector would be a potential market for photovoltaic energy converters if their installed operational cost can be lowered to the four dollars per square foot range.

Solar heating by thermal absorption in conjunction with photoelectricity generation offers some advantages in total installation cost reduction while providing both space heating and electricity for home use.[36] Studies of cost reduction possibilities in mass producing CdS-Cu_2S photovoltaic converters have indicated that if stability and efficiency problems traditionally associated with these cells can be eliminated they may

36. K. W. Boer, "Direct Energy Conversion for Terrestrial Use," Conference Record of the 9th IEEE Photovoltaic Specialists Conference, Silver Spring, Maryland, May 1972, pp. 351-358.

be produced and installed at a cost which would allow competition with utility power.[36,37]

The problems of selling a high technology item such as a residential solar power generation system in the multifaceted homeowner and new home construction markets are formidable. An optimist might estimate that the number of systems one could hope to sell under very favorable cost comparisons with alternative sources of electricity would be half the number of new houses built. If the rate of new housing starts is about two million per year[38] and each using solar energy required 1000 ft^2 of solar cells (10 kW peak power), this would amount to one billion ft^2 per year.

Mobile homes which are being produced at a rate of about one half million per year[38] may provide some market potential also. If the average mobile home roof could accommodate 400 ft^2 of solar converters, at 10% efficiency about 4 kW peak power (direct sunlight) could be supplied for air conditioning and other electrical usage. If photovoltaic converters were inexpensive enough a potential market for 200 million ft^2 per year might be expected, plus replacements.

---

37. J. O. Aaron and S. E. Isakoff, Third Conference on Large Scale Solar Energy Conversion for Terrestrial Use, Delaware, October 1971.

38. Survey of Current Business, U.S. Department of Commerce, 53, No. 12, December 1973.

Refrigerated truck-trailers, houseboats, recreational vehicles, and buses may provide a comparable market.[38] However, portable storage capacity or alternative generation would be needed to provide for non-sunlit hours because utility power would not be as readily available as it is to mobile homes.

In general, photo generation would be most acceptable on mobile vehicles which remain stationary for prolonged periods. Such vehicles are often equipped with large battery storage which is charged while the motive engine is in operation. While stationary, on water or in remote land areas, these batteries could be charged by photocells quietly without use of bulky and perhaps scarce fuel.

Another market for kiloWatt size electrical power systems which is expected to be very large in the future is in powering electric cars. Present prototypes offer speeds up to 80 miles per hour and performance comparable to a conventionally powered automobile, but a range of only 40-80 miles before a recharge of batteries is required.[23] With improvement of a power-to-weight and cost factors in batteries[39,40] one might expect that 30 horsepower (22 kW) "town cars"

---

39. Battelle Research Outlook, "New Frontiers in Energy Storage," J. McCallum and C. Faust, Battelle, Columbus, Volume 4, Number 1, 1972, p. 27.

40. "Ein Sonenelektrisch Angetriebener Kraft-Wagen," Radio u. Fernsehen, 10, 558 (1961).

will be commercialized in the next 20 years. Such automobiles may be expected to have about two hours storage capacity, about 40-50 kWhr. To supply 22 kW would require 2200 ft^2 of 10% efficient photovoltaic energy converters in direct sunlight. Hence it is not feasible to provide adequate direct power for cars by incorporating solar cells into its outer structure. However, if the average recharge of the battery system required about 25 kWhr it could be provided by an area of about 300 ft^2 (parking space) in a nine hour daylight period. One might envision electric car parking lots with photovoltaic electricity generating roofs. The same may be proposed for residential garage roofs. In both cases, utility power could be used for recharging in inclement weather and in the normally off-peak night time hours.

If one million electric cars were sold per year (less than 10% of automobile sales in the U.S.), and only 10% of these resulted in a demand for photovoltaic generation as described above, a market for 30 million ft^2/yr of photovoltaic converters would be produced. The cost of electricity produced in this way would have to be comparable to commercially available electricity converted to the proper dc voltage. Hence, the operational photoarray would have to cost about $4/$ft^2$ or less.

Other markets for kW-size photo generated electricity may include remote military installations, temporary construction sites where industrial generators are commonly used, off-shore oil production pumping stations, and pumping for irrigation.[41] The market size would be highly dependent upon location and cost of photovoltaic arrays. In some military locations the effective cost of fuel to provide power to generators may range up to \$10/gallon.[42] Windmills for pumping water cost about \$1000-2000 per horsepower. The solar pump could be expected to operate at a load factor at least as great as windmills, hence it would be competitive at a cost of about \$1500/kW peak power or \$15/ft^2 for 10% efficient photo systems. An area 10 x 7.5 ft could provide one horsepower peak pumping capacity in remote cattle grazing lands for providing water for livestock. Storage capacity for inclement or darkness hours would be present in the form of water tank reservoirs as are now associated with windmills. For the latter application, one might expect eight thousand units per year to be a maximum market in the U.S.[42] This would amount to about one half million square feet of solar cells per year, plus replacements. Other areas of the world with less rural

---

41. N. Lidorenko, F. Nabiullin, A. Landsman, B. Tarnizhevskiy, and Ye. Gertsik, Heliotechnology, 3, 1 (1966).

42. A. Zarem and D. Erway, "Introduction to the Utilization of Solar Energy," McGraw-Hill Book Company, New York, 1963, pp. 211-238.

electrification may provide a much greater market potential.

Watt - The market for small sources of photoelectricity can be estimated by analyzing the market for batteries. In 1966 the energy accumulating capacity of batteries was about 62 million kWhr. This was 0.007% of the 881,000 million generated in the U.S. that year.[39] Because battery and electricity usage have increased at about the same rate this percentage should still be approximately the same. Then the present battery storage capacity should be about 140 million kWhr. If this were to cost 1.8 cents/kWhr, the average price of commercial electricity, it would have a value of about $2.5 million. However, battery power costs up to $500/kWhr, depending upon the type of battery. At present it is a $1.3 billion business.[43] For the advantages of convenience and portability consumers are willing to pay thousands of times the cost of utility power in small and rather specific applications.

In estimating how much of this market could be served by photovoltaic systems it is instructive to define where batteries are used. Secondary batteries

---

43. D. Shearer, Jr. and R. Farrell, "Battery Power," Long Range Planning Service, Stanford Research Institute, Menlo Park, Calif. 1969, Report No. 396.

(rechargeable) account for about two thirds of the dollar market and about 90% of these are lead-acid types of which nearly all are used for starting, lighting, and ignition and in industrial fork lift trucks. The other 10% of secondary batteries, i.e., about $90 million, are nickel-cadmium, silver-zinc, silver-cadmium, and nickel-iron, most of which are used in rechargeable toothbrushes, razors, and similar applications. It would appear that the preponderance of the rechargeable battery market is served well by engine powered generators in automobiles, aircraft, and boats, and by the electrical utilities in the indoor use areas. Hence, photovoltaic generation would not offer any particular advantage nor command any higher price.

Smaller markets for photoelectricity to recharge batteries lie in reserve batteries in remote locations. Photocells can be used to maintain charge in batteries used on boats and portable or remote signal lights such as used to mark tall structures, road work, seaways, and offshore oil pumping stations.[44] Presently photovoltaic energy converters are used to provide power and recharge for forest observation sites,[45]

---

44. Brochure for Solar Power Module 1002, Solar Power Corporation, Braintree, Mass. 1974.

45. W. Hasbash, Sun at Work, 10, 3 (1965).

lighthouses,[46] repeater stations, buoys, and other navigation, communications, and signaling applications.[47] Of course they are used nearly exclusively to power earth satellites and a yearly expenditure of about $15 million is expected for this purpose through the 1970s.[48] The total market for reserve batteries is about $30 million per year.[43]

The rest of the battery market is in primary cells (single use) such as the carbon-zinc flashlight type, alkaline-manganese, and magnesium-anode type. The present annual sales are about $400 million.[43] These are typically used in radios, toys, lanterns, portable communication equipment, cameras, clocks, hearing aids, phonographs, alarm systems, transmitters,...etc. Because most uses are on-demand, photo generated power would necessarily have to be accompanied by secondary storage cells such as nickel-cadmium batteries. Uses are randomly indoors, outdoors, day and night. The cost of the combined solar cell and secondary battery power system would have to be competitive with that of primary batteries, i.e., about an average of $20/kWhr.[39] Although solar powered flashlights, sewing machines,

---

46. S. Hirai and M. Kobayashi, "Unattended Lighthouse Using Solar Batteries as Power Source," U.S. Coast Guard, Washington, D. C., 1960, Report 4-6-2, AD-242039.

47. C. K. Sterkin, "Terrestrial Applications of Solar Cells," Literature Search No. 895, Jet Propulsion Laboratory, 1968.

48. See reference 19, Chapter 4.

toys, telephones, electric fence chargers, clocks, radios, headsets, alarm systems, highway emergency call boxes, and the like have been manufactured, the market is very segmented and difficult to quantify. It is here, however, where the cost of power is high (and in remote area usage) that photovoltaic converters will gain increasing acceptance as their price decreases in the future. If a 10% penetration of this market were forthcoming by the year 2000 it would be about a $140 million market assuming a rather conservative 5% annual growth rate.[43]

Chapter 6

BUSINESS OPPORTUNITIES AND THE DEVELOPMENT OF PHOTOVOLTAIC ENERGY CONVERSION SYSTEMS

As illustrated in the preceding chapters the business opportunities in the manufacture and sale of photovoltaic materials are a strong function of the cost of the photoelectricity produced by light interaction with these materials and appropriate storage for on-demand use. This cost is inextricably tied to the cost of manufacture of photocells which is presently about $20 per peak power Watt for terrestrial uses, i.e., $200/ft^2. This is about 50-100 times too expensive to yield electricity at a cost competitive with conventionally produced utility power. However, photovoltaic energy conversion is relatively new and has not had the attention of people knowledgeable in mass production techniques nor has its development been supported by governmental funding to any great extent. Hence, cost reductions of the order mentioned may be in the realm of possibility. With the present concern for energy supplies as well as for the environment, incentives are formulating which portend more research and development in this technology.

The U.S. federal government spent about an average of about $100,000 annually on solar energy research from 1950 to 1970. In searching the literature it may be discerned that the European countries, Japan, Australia, and the U.S.S.R., have also funded solar energy research, but to a rather halting and modest extent. Since 1970 the United States, through the National Science Foundation, has begun to increase funding. The actual and estimated expenditures are illustrated in Table 3.

Table 3* - NSF Solar Energy Budget ($\$ \times 10^6$)

Year	Photovoltaic Conversion	Total Solar
1971	negligible	1.20
1972	0.43	1.66
1973	0.89	3.96
1974	2.40	13.20
1975	8.0	49.0

*Derived from references 1,2

Recently, special legislation has been considered to increase this support. The Solar Heating and Cooling Demonstration Act would provide $50 million to demonstrate solar heating in homes, apartments, offices, factories, and schools. Some of the construction development would be applicable to providing photoelectricity in combination with thermal conditioning.

Because business opportunities in the past have not provided the adequate incentive for private development

1. W. A. Shumann, Aviation Week and Space Technology, January 14, 56 (1974).

2. A. I. Rosenblatt, Electronics, April 4, 99 (1974).

of photovoltaic converters at an economically competitive electricity cost, the governmental funding of research and development is crucial to the advancement of photovoltaic energy conversion (see Figure 23). If it can induce a major cost reduction from advanced technology, more entrepreneurial efforts will be made with further cost reductions from mass production and concomitant learning processes. One can estimate somewhat how much effect governmental support may have by referring to learned opinions. The Ad Hoc Panel on Solar Cell Efficiency, National Research Council,[3] estimated that $5 million applied appropriately would increase the efficiency of silicon solar cells by about 50%; $20 million by about 100%. Hence, the effective price of silicon cells might be halved by a $20 million funding. In reference 4 it was estimated that an effective cost reduction of eight times could be achieved by concentrating light on cooled silicon cells with present technology. Thus, a possibility of cost reduction of 16 times has been outlined. If this were achieved, another factor of 3-10 would be required to make photoelectricity cost competitive with utility power.

On the basis of the last ten years, which saw silicon wafer cost decrease by 50% for each order of

---

3. See reference 19, Chapter 4.

4. See reference 52, Chapter 4.

magnitude in manufacturing volume one might expect some further reduction.[5] Basic materials cost, the cost of electricity for the silicon growth process, and inherent wastes in the manufacturing would dictate that costs of the finished cell will not fall below \$1000/kW peak power, i.e., \$10/ft^2. However, this reduction combined with efficiency gains and concentrators mentioned above may yield photovoltaic converters with a cost of \$200-400/kW which would allow extensive utilization.

Some authors estimate[6] that the cost of 18% efficiency silicon solar cell arrays can be reduced to yield electricity at a cost of \$558/kW to \$1720/kW in space applications (the same array would cost about 25% more per peak power kW in terrestrial situations, plus weather protection costs). The lower value has a "25% confidence level" while the higher is 75%. The "normal extrapolation of today's technology to the 1980s" using historical experience of the space power business growth over the decade of the 1960s yields an estimated cost of \$14-37/ft^2 for arrays with a concentration factor of three.[6] One might expect that water

5. C. Currin, K. Ling, E. Ralph, W. Smith, and R. Stirn, "Feasibility of Low Cost Silicon Solar Cells," Conference Record of the 9th IEEE Photovoltaic Specialists Conference, Silver Spring, Maryland, May 1972, pp. 363-369.

6. E. Ralph and F. Benning, "The Role of Solar Cell Technology in the Satellite Power Station," _ibid._, pp. 370-381.

cooling and higher concentration factors may decrease this estimated cost for terrestrial applications by a factor of about three.

New methods for growing silicon crystals may also reduce the cost considerably. It has been estimated by several experts in silicon technology[5] that ribbon growing of silicon as in the edge-defined, film-fed growth[2,7,8,9] may reduce the cost of finished silicon solar cells to $3.50/ft^2. This would yield a cell cost of about $350/peak power kW for 10% efficiency cells. If the efficiency were increased the cost would be correspondingly less. The main technical parameter limiting this method is that the material used as a die is often corroded with resultant deterioration of the silicon purity. This would not appear to be a long term obstacle.

P. Glaser has predicted[10] that the cost of electricity from solar cells will drop from the present $20,000/kW to $5,000 in 1979 and $1,000 by 1984. The

---

7. H. Bates, F. Cocks, and A. Mlavsky, "The Edge-Defined, Film-Fed Growth (EFG) of Silicon Single Crystal Ribbon for Solar Cell Applications," op. cit., pp. 386-387.

8. Technology Trends, Optical Spectra, September, 30 (1974).

9. Technology, Chem. Eng. News, July 29, 16 (1974).

10. See reference 4, Chapter 5.

considered opinions of several experts surveyed in this research (see preface) generally were to the effect that with "proper" funding the cost could be lowered to $1,000/kW in ten years and possibly to $200/kW by the end of the century. A recent Delphi study with 63 participants carried out by A. D. Little[11] indicated nearly the same; but $500/kW in 25 years. The present energy concerns and recent increases in federal funding (see Table 3) may mean that "proper" funding is imminent. The cost reduction for silicon solar cells as a function of time is estimated as in Figure 24.

The development of cost-effective photovoltaic converters could be greatly influenced by other types of photovoltaic materials such as the CdS-Cu_2S. Because of less historical emphasis on this material it is very difficult to estimate its future development. Although its manufactured cost can be much less than silicon cells presently, it does not have the efficiency nor the stability of the latter. Recently, it was estimated that these materials when properly encapsulated may in the terrestrial environment last up to ten years.[12] However, hard data of this achievement are yet to be published. These cells presently have

---

11. P. Glaser, J. Berkowitz, R. LaRose, A. D. Little, Inc., Cambridge, Mass., Private Communication, 1974.

12. See reference 36, Chapter 5.

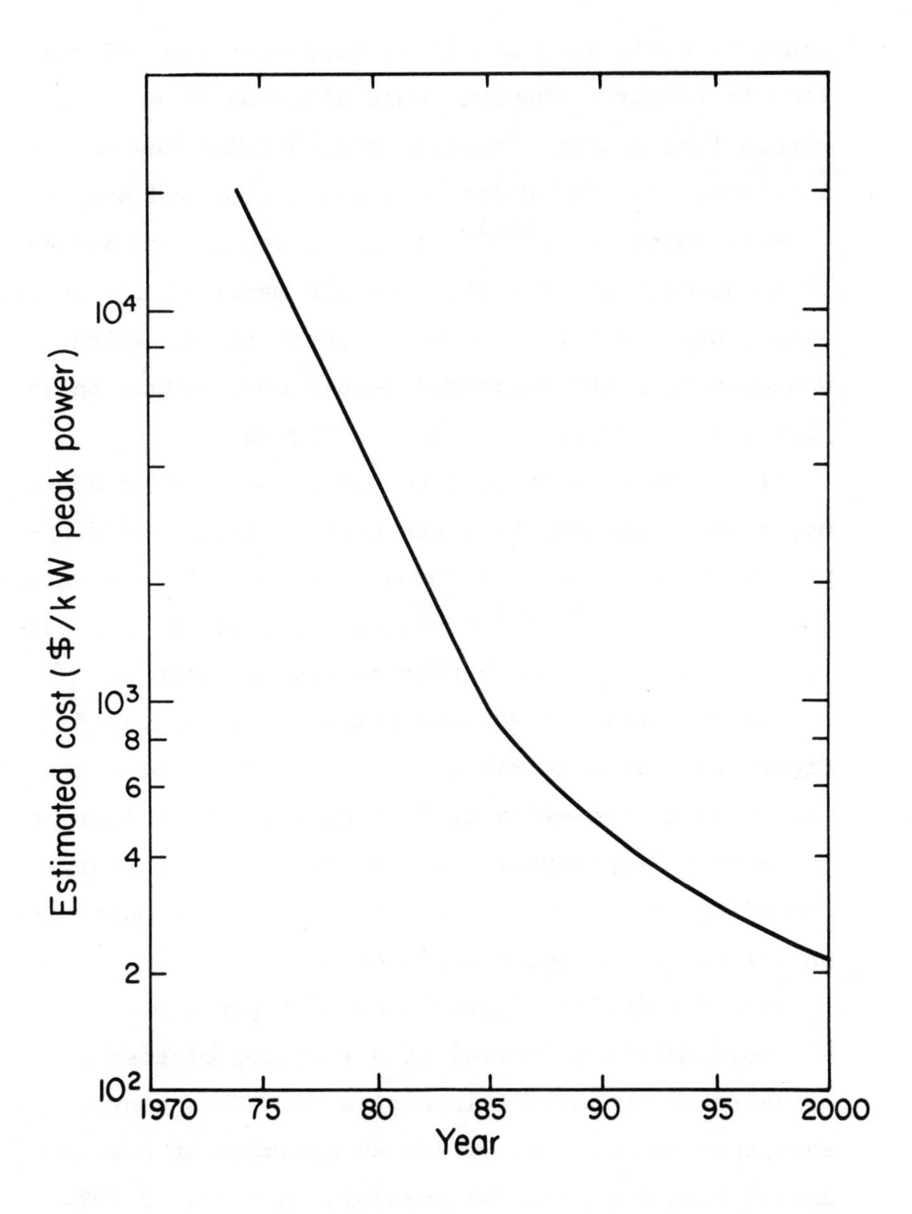

Figure 24. Forecast of the Cost of Silicon Solar Cells as a Function of Time.

about 5% efficiency and it is suggested that $5 million in research support would allow an 8% efficiency within five years. Properly sealed cells may cost as little as $1-2/ft^2 using mass production and state-of-the-art technology.[12,13] It is reasonable to assume that CdS-Cu_2S and new photovoltaic material discoveries will provide alternative development routes which would sustain the estimated future cost reduction in photoelectricity represented in Figure 24.

If the technological developments discussed above occur as predicted, business opportunities would appear very favorable. A summary of maximum 1974 market potentials in ft^2 of 10% efficiency photovoltaic converters (reasoned in Chapter 5) for operational installed arrays at various prices is illustrated in Figure 25. If electricity and battery use predictions are correct, the estimate in Figure 25 would increase by about 6.9% per year for the rest of the century. From Figures 24 and 25 one can estimate the potential annual sales of photovoltaic converters.

The convolution of predicted cost per square foot and 1974 market potential as a function of cost is illustrated by curve A in Figure 26. This convolution as a function of time yields an estimate of how much annual business might be possible in terms of 1974 demand and dollar value. However, in a new business where production capacity would have to be built, a

---

13. See reference 37, Chapter 5.

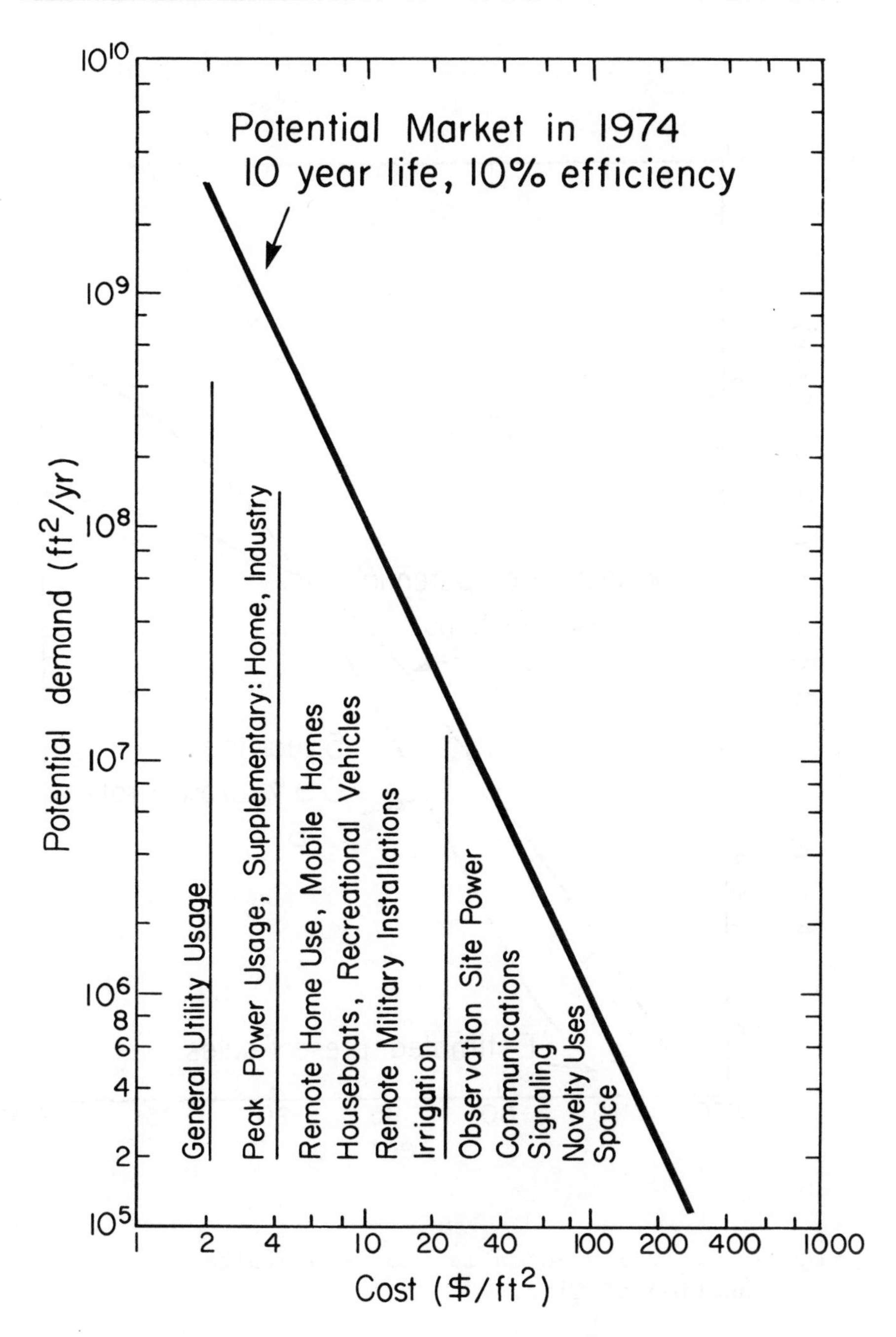

Figure 25. Demand for Photovoltaic Arrays as a Function of Cost.

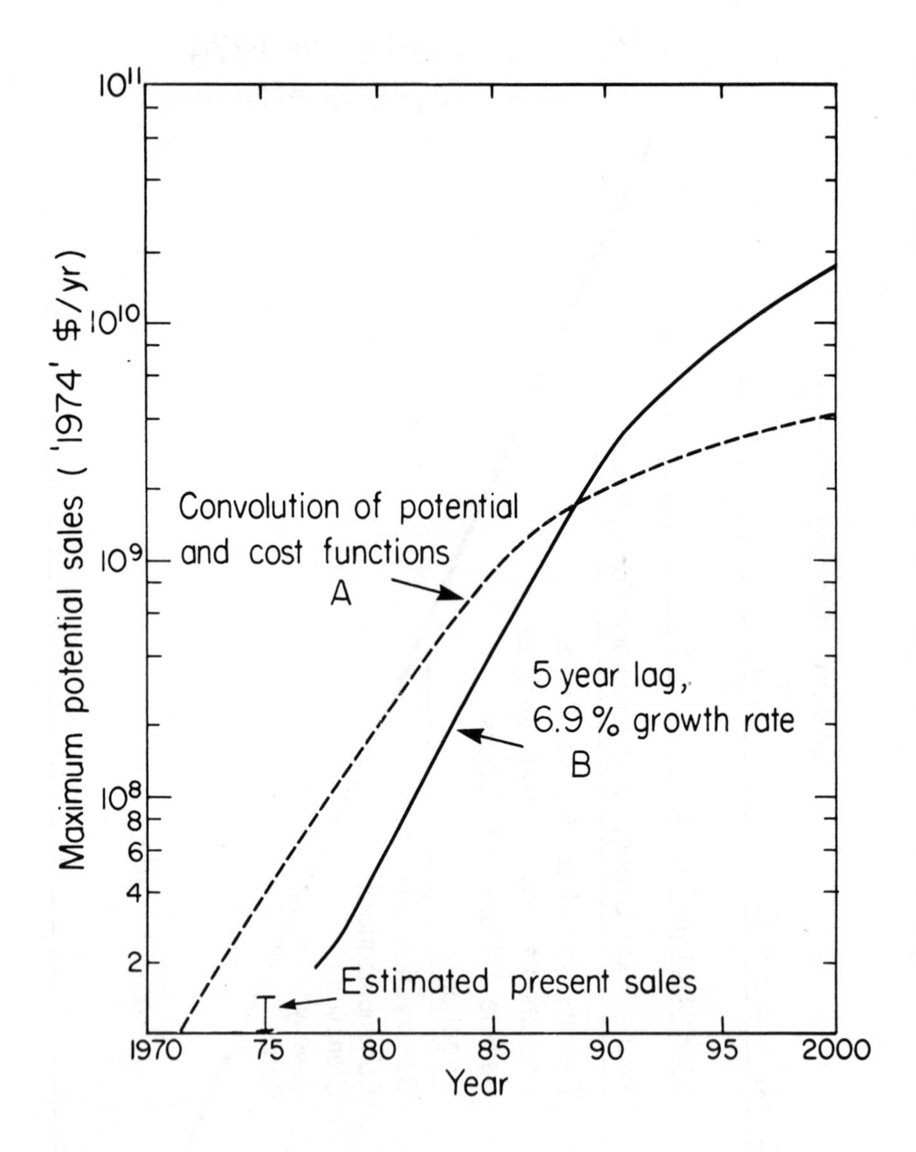

Figure 26. Sales Potential for Photovoltaic Arrays as a Function of Time.

certain time lag should be allowed. Growth in demand as a function of time must be considered also. Curve B in Figure 26 represents the estimated, annual sales potential in 1974 dollars if: (1) the cost reduction illustrated in Figure 24 is achieved; (2) the potential demand function shown in Figure 25 grows evenly in all cost sectors at an annual rate[14] of 6.9%; and (3) there is a five year lag of supply behind the demand at a given price.

It might appear from Figure 26 curve B that the potential business is very great, reaching $15 billion by the year 2000. This may seem somewhat misleading in that lag times in converting from alternative conventional sources of electricity to photovoltaic electricity may be very long. With lifetimes of 20-50 years such facilities would not be abandoned unless the associated variable costs of electricity produced exceeded the cost of photoelectricity—an unlikely situation. On the other hand, with a high electricity demand growth rate (about 6.9%/yr)[14] the opportunity for introduction of photovoltaic converters into new peak power facilities and in the general utility power field may be very high by the end of the century. Let us add perspective by considering the possible demand for photovoltaic converters in the year 2000 if they were priced at two dollars per square foot, operational

---

14. See reference 4, Chapter 1.

and with required storage. A relatively conservative estimate of the yearly electricity demand is 7×10^{12} kWhr.[15] If a 10% efficient converter produced an average of 1.7 W/ft^2 for 8766 hrs in the year it would produce 14.9 kWhr of electricity. Hence it would take 4.7×10^{11} square feet to supply the total. If we assume that only 10% of the production to meet this demand were by photovoltaics, 4.7×10^{10} square feet would be required. Let us next assume that there is a 10 year lifetime of the photovoltaic arrays so the replacement rate is 4.7×10^9 square feet per year. Thus a business of 4.7×10^9 ft^2/yr times two dollars per square foot, or 9.6 billion dollars, would be established just for replacing 1% of the capacity to meet the electricity demand. One can see that if the demand increases by several percent per year and a significant fraction of the new capacity to meet that demand is by photovoltaics a multibillion dollar business is within the realm of possibility.

By inspection of Figures 24, 25, and 26, it is obvious that technological advancement with major concomitant cost reductions is imperative for major sales of photovoltaic converters. If the predicted cost reductions do not materialize, it is doubtful that photoelectricity will be in general use by the end of the century. Even if such reductions in photovoltaic

---

15. E. Hudson and D. Jorgenson, Bell J. Econ. Mgmt. Sci., 5, 461 (1974).

converter costs occur, concomitant major cost reductions in electrical storage will be necessary for cost-competitive on-demand usage applications. The storage cost reduction required would most likely occur in large utility operations where pumped-water storage or H_2 storage could be utilized effectively.

If solar energy is to be used to significantly enhance the supply of electricity it is clear that research in photovoltaic materials and manufacturing processes will have to provide the pathway. As illustrated in Figure 23, funding of this research will probably have to come from governmental support and/or far-sighted support within industry. If cost-competitive photovoltaic converters become a reality, there would be considerable opportunity for future product and market development. Product refinement could take various forms of higher reliability, more versatile modules, enhanced styling, greater ease of installation, and perhaps most importantly, increased efficiency. Although an increase in efficiency of 100% may almost bring the present state-of-the-art efficiency to the theoretical limit, small increments to this aim would mean major differences in areas needed and potential uses of photovoltaic arrays. For example, a doubling of efficiency would mean the average house roof area may support electrical heating of the residence as well as the other electrical needs. Visual degradation of the environment, if any, would be reduced by using smaller arrays.

The potential for development of world markets may also be very favorable for the future sales of photovoltaic converters. A recent report by the Ad Hoc Advisory Panel of the Board on Science and Technology for International Development[16] indicated that the probable growth in electricity usage in developing countries will be 10-15% per year if it can be made available. It is expected that as the gross national product per person increases the electrical energy usage per capita will increase also. Indeed, if these data for various countries are placed on a log-log plot, a linear graphic relationship is found.[17] It is also generally true that the developing countries are in, or adjacent to the tropics and have favorable solar radiation conditions, while conventional energy sources are not readily available.[16] They are often characterized by arid climates, dispersed and inaccessible populations, readily available labor, and little investment capital. Under such conditions small and disperse photovoltaic power stations may be preferable to larger conventional stations which require a considerable power transmission investment.

16. Ad Hoc Advisory Panel of the Board on Science and Technology for International Development, "Solar Energy in Developing Countries: Perspectives and Prospects," National Academy of Sciences, Washington, D. C., 1972.

17. InterTechnology Corporation, "The U.S. Energy Problem, Volume I, Summary," NITS, U.S. Department of Commerce, Springfield, Va., 1971, p. 28.

Summary and Predictions

The availability of terrestrial sunlight is abundantly sufficient to meet the energy needs of the United States and most parts of the world. The advantages of its long term and widespread availability, and pollution-free utilization are offset somewhat by its short term fluctuations and the capital investment needed to provide for its collection and storage in a form acceptable for use during unilluminated hours.

The scientific principles upon which the photovoltaic effect is based allow that up to about 25% of collected sunlight can be converted directly into electricity at ambient temperatures. The state-of-the-art permits about 12% efficiency using silicon devices which have at least 10 years life and negligible degradation. The cost of silicon cells is 50-100 times higher than that necessary to compete economically with conventional means of generating electricity. The cost of electrical storage necessary to provide on-demand electricity independent of other energy sources is about 2-5 times greater than required for economic viability in general usage.

The expected research funding and technological development of photovoltaic energy converters portends that photoelectricity will become cost competitive for peak-power and general supplementary use by about 1990 and for general power production by 2000. Photovoltaic converters, which are most advantageous for remote hard-to-service locations such as space, will first gain acceptance in novelty uses, then in communications, water pumping, recreational vehicles, mobile and remote homes, supplementary peak-power generation, and finally in general power generation, as their cost decreases.

With the projected technological development, the potential market for 10% efficiency photovoltaic energy converters should increase from about 10^5 ft^2/yr presently, to about 10^9-10^{10} ft^2/yr in 2000. The high growth rate in electrical usage will allow rapid expansion of the photovoltaic converter industry in the last decade of this century leading to a possible multibillion dollar business by 2000. By assuming various cost reductions in photovoltaic arrays and associated storage one can schematically illustrate the range of business opportunities which may develop. Figure 27 represents my prognistication of the range of business opportunities which will be present 25 years in the future. This was derived by making considered judgments of the probabilities of cost reductions based upon the information gathered in this study, relating costs to potential markets as illustrated in Figure 25, and compounding a 6.9% yearly growth rate in electricity usage. The reader may wish to choose different probabilities and costs dependent upon his best judgment.

In the final analysis it is demand which will stimulate the innovation required to make photovoltaic

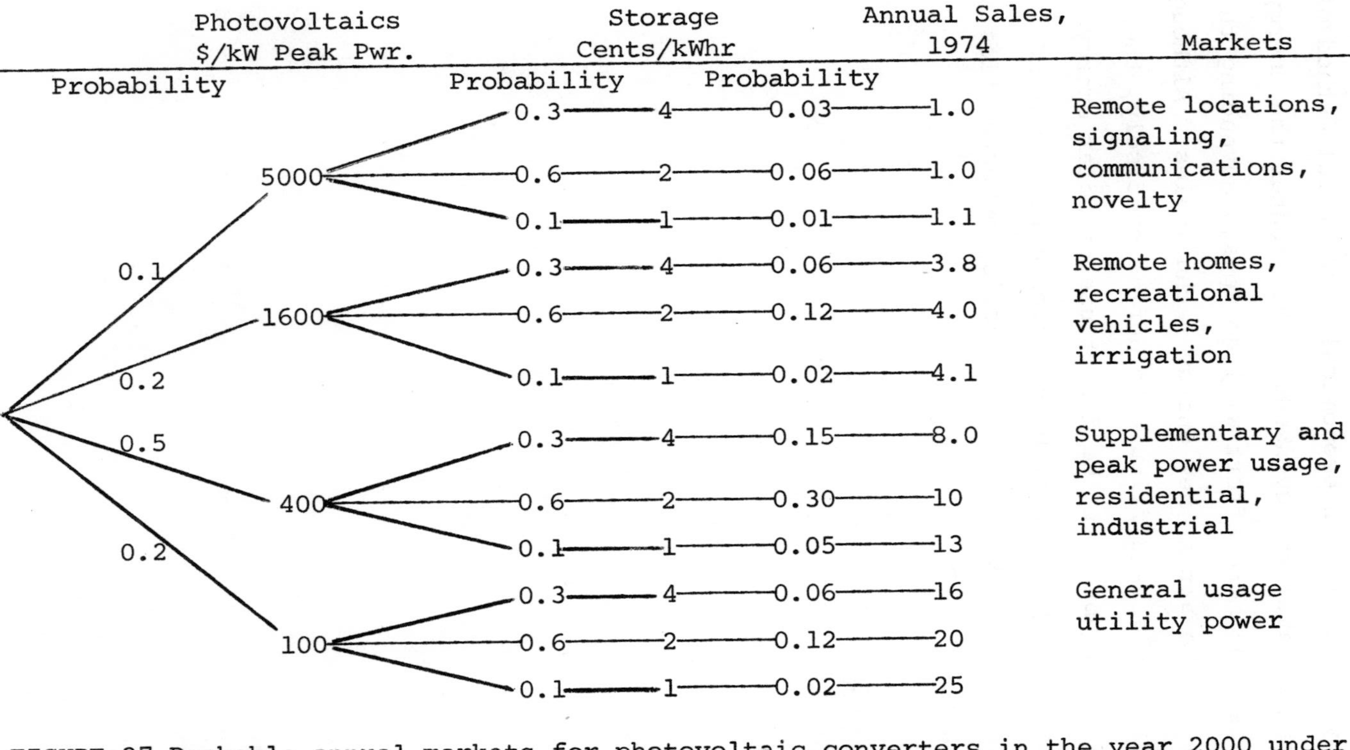

FIGURE 27. Probable annual markets for photovoltaic converters in the year 2000 under projected costs. Present costs: Photovoltaics $20,000/kW, storage 6 cents/kWhr.

energy conversion an economically practical method of meeting general energy needs. The greater the energy "crisis" the more effort will go toward advancing the state-of-the-art in new energy technologies. Although the timetable is difficult to define precisely, there is little doubt that the direct use of solar energy for heat and power will become much more prevalent within this century. Entrepreneurial opportunities will abound.

BIBLIOGRAPHY

Aaron, J. O.; and Isakoff, S. E., Third Conference on Large Scale Solar Energy Conversion for Terrestrial Use, Delaware, October 1971.

Abetti, G., "The Sun," Faber and Faber, London, 1963, p. 273, translated by; J. B. Sidgwick.

Abrahamsohn, U.S. Patent 3,376,163; April 2, 1968.

Ad Hoc Advisory Panel of the Board on Science and Technology for International Development, "Solar Energy in Developing Countries: Perspectives and Prospects," National Academy of Sciences, Washington, D. C. 1972.

Ad Hoc Panel on Solar Cell Efficiency, P. Rappaport, Chairman, "Solar Cells; Outlook for Improved Efficiency," National Academy of Sciences, Washington, D. C., 1972.

Allen, C. W., Quart. J. Roy. Met. Soc., 84, 307 (1958).

Altman, M., "Elements of Solid State Energy Conversion," Van Nostrand Reinhold Co., New York, 1969, pp. 240-263.

Angrist, S. W., "Direct Energy Conversion," Allyn and Bacon, Inc., Boston, 1965, p. 326.

Associated Universities, Inc., Report "Reference Energy Systems and Resource Data for Use in the Assessment of Energy Technologies," April 1972. Report to U.S. Office of Science and Technology, under Contract OST-30; Document AET-8.

Atomic Energy Commission, "Nuclear Power Growth 1971-1985," Washington, December 1971.

Atzei, A.; Capart, J.; Crabb, R.; Heffels, K.; and Seibert, G., "Improved Antireflection Coatings on Silicon Solar Cells," Solar Cells, Gordon and Breach Science Publishers, New York 1971, pp. 349-362.

Backus, C. E., "A Solar-Electric Residential Power System," Seventh Intersociety Energy Conversion Engineering Conference, Conference Proceedings, September 1972, pp. 704-711.

Baker, B., Fuel Cell Systems-II, "Advances in Chemistry Series," Edited by R. F. Gould, American Chemical Society Publications, Washington, D. C. 1969.

Bates, H.; Cocks, F.; and Mlavsky, A., Conference Record of the 9th IEEE Photovoltaic Specialists Conference, Silver Spring, Maryland, May 1972, pp. 386-387.

Battelle Research Outlook; "Our Energy Supply and Its Future," Editor; A. B. Westerman, Battelle, Columbus, Volume 4, Number 1, 1972, p. 3.

Battelle Research Outlook, "New Frontiers in Energy Storage," J. McCallum and C. Faust, Battelle, Columbus, Volume 4, Number 1, 1972, p. 27.

Beauzee, C., U.S. Patent 3,261,074; July 19, 1966.

Beckman, W.; Schoffer, P.; Hartman, W., Jr.; and Lof, G., Solar Energy, 10, No. 3, 133 (1966).

Berman, Dr. Elliot, President, Solar Power Corporation, Braintree, Massachusetts, Private Communication, February 1974.

Berman, P. A., Solar Energy, 11, No. 3-4, 180 (1967).

Bernatowicz, D.; and Brandhorst, H., Jr., "The Degradation of Cu_2S-CdS Thin Film Solar Cells Under Simulated Orbital Conditions," Conference Record of the 8th IEEE Photovoltaic Specialists Conference, Seattle, Washington, August 1970, pp. 24-29.

Boer, K. W., Chem. Eng. News, January 29, 12 (1973).

Boer, K. W., Boston Globe, January 20, 1974, PA-2, Reported by J. Thomas.

Boer, K. W., "Direct Energy Conversion for Terrestrial Use," Conference Record of the 9th IEEE Photovoltaic Specialists Conference, Silver Spring, Maryland, May 1972, pp. 351-258.

Boll, R.; and Bhada, R., Energy Conversion, 8, No. 1, 3 (1968).

British Petroleum Company, "1972 Statistical Review of the World Oil Industry."

Brown, W. C., "Microwave Power Transmission in the Satellite Solar Power Station System," Raytheon Company Technical Report ER72-4038, 1972.

Bruckner, A., II; Fabrycky, W.; and Shamblin, J., IEEE Spectrum, April, 101 (1968).

Cheban, A.; Negreskul, V.; Oush, P.; Gorchak, L.; Unguranu, G.; and Smirnov, V., Geliotekhnika, 8, No. 1, 30 (1972). Applied Solar Energy, 8, No. 1-2, 23 (1973).

Cherry, W. R., Proceedings of the 13th Annual Power Sources Conference, Atlantic City, May 1959, pp. 62-66.

Cherry, W. R., "A Concept for Generating Commercial Electrical Power from Sunlight," Conference Record of the 8th IEEE Photovoltaic Specialists Conference, Seattle, August 1970, pp. 331-337.

Crabb, R., "Status Report on Thin Silicon Solar Cells for Large Flexible Arrays," Solar Cells, Gordon and Breach Science Publishers, N. Y., 1971, pp. 35-50.

Crossley, P.; Noel, G.; and Wolf, M., NASA Report NASW-1427.

Culberson, O. L., "The Consumption of Electricity in the United States," Oak Ridge National Laboratory, 1971, ORNL-NSF-EP-5.

Currin, C; Ling, K.; Ralph, E.; Smith, W.; and Stirn, R., "Feasibility of Low Cost Silicon Solar Cells," Conference Record of the 9th IEEE Photovoltaic Specialists Conference, Silver Spring, Maryland, May 1972, pp. 363-369.

Daniels, F., "Energy Storage Problems," Solar Energy, 6, No. 3, 78 (1962).

David, J.; Martinuzzi, S.; Cabane-Brouty, F.; Sorbier, J.; Mathieu, J.; Roman, J.; and Bretzner, J., "Structure of CdS-Cu_2S Heterojunction Layers,: Solar Cells, Gordon and Breach Science Publishers, New York, 1971, pp. 81-94.

Denton, J.; and Herwig, L., Proceedings of the 25th Power Sources Symposium, May 1972, p. 137.

Diermendjian, D.; and Sekera, Z., Tellus, 6, 382 (1954).

Dorokhina, T.; Zaytseva, A.; Kagan, M.; Polisan, A.; and Kholev, B., Geliotekhnika, 9, No. 2, 6 (1973). Applied Solar Energy, 9, No. 1-2, 50 (1974).

Duchesneau, T. D., Federal Trade Commission Economic Report, "Interfuel Substitutability in the Electric Utility Sector of the U.S. Economy," U.S. Government Printing Office, Washington, D. C., 1972.

Edison Electric Institute Statistical Yearbook of the Electric Utility Industry for 1972, New York, No. 40, Publication 73-13, November 1973.

Edison Electric Institute, A report of National Economic Research Associates to the Edison Electric Institute, "Fuels for the Electric Utility Industry 1971-1985," New York, 1972, EEI Pub. No. 72-27.

Egorova, I. V., F.T.P., 2, 319 (1968).

Elliott, J. F., "Photovoltaic Energy Conversion," in Direct Energy Conversion, Edited by G. W. Sutton, McGraw-Hill Co., New York, 1966, pp. 1-37.

Federal Power Commission, "1970 National Power Survey," Washington, D. C., Volume IV, 1971, p. 3.

Federal Power Commission, "1970 National Power Survey," Washington, D. C., Volume I, 1971, p. 19.

Forrester, J. W., "Principles of Systems," Wright-Allen Press, Cambridge, Mass., 1972, Second Preliminary Edition.

Forrester, J. W., "Industrial Dynamics," M.I.T. Press, Cambridge, Mass., 1961.

Gavrilova, N., Applied Solar Energy, 8, No. 5-6, 68 (1973).

Ghosh, A.; and Feng, T., J. Appl. Phys., 44, 2781 (1973).

Ghosh, A.; Morel, D.; Feng, T.; Shaw, R.; and Rowe, C., J. Appl. Phys., 45, 230 (1974).

Glaser, P. E., "Space Resources to Benefit the Earth," Third Conference on Planetology and Space Mission Planning, The New York Academy of Sciences, October 1970.

Glaser, P., Testimony before the House of Representatives, reported by J. Thomas, Boston Globe, January 20, 1974, p. A-3.

Glaser, P. E., "The Future of Power from the Sun," Intersociety Energy Conversion Engineering Conference 1968 Record, August 1968, p. 98.

Glaser, P.; Berkowitz, J.; LaRose, R., A. D. Little, Inc., Cambridge Mass., Private Communication 1974.

Gold, R., "Current Status of GaAs Solar Cells," Transcript of Photovoltaic Specialists Conference, Vol. 1, Photovoltaic Materials, Devices and Radiation Damage Effects, DDC No. AD412819, July 1963.

Goody, R. M., "Atmospheric Radiation," Clarendon Press, Oxford, 1964, pp. 417-426.

Gorski, D. A., U.S. Patent 3,186,874; June 1, 1965.

Griffin, J., Bell J. Econ. Mgmt. Sci., 5, 515 (1974).

Grubb, W.; and Niedrach, L., "Fuel Cells," in Direct Energy Conversion, Edited by G. W. Sutton, McGraw-Hill Co., New York, 1966, pp. 39-104.

Guillien, M.; Leitz, P.; Marchal, G.; and Palz, W., Solar Cells, Gordon and Breach Science Publishers, New York 1971, pp. 207-214.

Halsted, R., J. Appl. Phys., 28, 1131 (1957).

Hasbash, W., Sun at Work, 10, 3 (1965).

Haworth, L. J., "Assessment of Energy Technologies Report on Step 1," Associated Universities, Inc., Upton, N.Y., 1971, p. 103.

Henderson, S. T., "Daylight and Its Spectrum," American Elsevier Publishing Co., New York, 1970.

Hirai, S.; and Kobayashi, M., "Unattended Lighthouse Using Solar Batteries as Power Source," U.S. Coast Guard, Washington, D. C., 1960, Report 4-6-2, AD-242039.

Hudson, E.; and Jorgenson, D., Bell J. Econ. Mgmt. Sci., 5, 461 (1974).

Huth, J., U.S. Patent 2,993,945; July 25, 1961.

Iles, P.; and Ross, B., U.S. Patent 3,361,594, January 2, 1968.

InterTechnology Corporation, "The U.S. Energy Problem, Volume I, Summary," NITS, U.S. Department of Commerce, Springfield, Va., 1971, p. 28.

ITC Report C645, "The U.S. Energy Problem," Volume I Summary Volume," InterTechnology Corporation, Warrenton, Virginia, 1971.

Joint Committee on Atomic Energy, "Certain Background Information for Consideration When Evaluating the National Energy Dilemma," U.S. Printing Office, Washington, 1973.

Kastens, M. L., "The Economics of Solar Energy," Introduction to the Utilization of Solar Energy," Edited by A. Zarem and D. Erway, McGraw-Hill Book Co., New York, 1963, pp. 211-238.

Kaye, S., "Drift Field Solar Cells," Transcript of the Photovoltaic Specialists Conference., Vol. 1, Photovoltaics Materials, Devices and Radiation Damage Effects, DDC No. AD412819, Sec. A-6, July 1963.

Komashchenko, V.; Marchenko, A.; and Fedorus, G., Poluprovodnikovoya Tekhnika; Makroelectronika, No. 4, 112 (1972). AD-756594, NTIS, U.S. Department of Commerce, Springfield, Va.

Landsberg, H.; Lippmann, H.; Paffen, Kh.; and Troll, C., "World Maps of Climatology," Springer-Verlag, New York, 1965, Edition 2.

Lapin, E.; Ernest, A.; and Sollow, P., U.S. Patent 3,427,200; February 11, 1969.

Lawrence Livermore Laboratory, "Energy: Uses, Sources, and Issues," UCRL-51221 May 30, 1972.

Lebrun, J.; and Bessonneau, G., "New Work on CdTe Thin Film Solar Cells," Solar Cells, Gordon and Breach Science Publishers, New York 1971, pp. 201-206.

Lidorenko, N. S., Elektrotekhnika, No. 2, 1 (1967).

Lidorenko, N. S., Geliotekhnika, 7, No. 2, 52 (1970).

Lidorenko, N.; Nabiullin, F.; Landsman, A.; Tarnizhevskiy, B.; and Gertsik, Ye., Heliotechnology, 3, 1 (1966).

Lindmayer, J., Proceedings of the 9th IEEE Photovoltaic Specialists Conference, Maryland, May 1972.

Lof, G.O.G., "Engineering and Economic Problems in the Production of Electric Power from Solar Energy," World Power Conference on Energy, Rio de Janiero, 1954.

Loferski, J., J. Appl. Phys., 27, 777 (1956).

Lyons, L.; and Newman, O., Australian J. Chem., 24, 13 (1973).

Mantell, C. L., "Batteries and Energy Systems," McGraw-Hill Book Co., New York, 1970.

Marchenko, A. I.; and Fedorus, G. A., U.F.Zh., 12, 1392 (1967).

Meinel, A. B.: and Meinel, M. P., Bull. At. Scientists, 27, No. 8, 32 (1971). Also: "Solar Energy for the Terrestrial Generation of Electricity," Hearing before the Subcommittee on Energy, June 5, 1973, U.S. Printing Office, Washington, D. C.

Miller, S., U.S. Patent 3,081,370; March 12, 1963.

M.I.T. Energy Laboratory and M.I.T. Lincoln Laboratory, "Proposal for Solar-Powered Total Energy Systems for Army Bases," Massachusetts Institute of Technology, July 1973.

Mockovciak, J., Jr., "A Systems Engineering Overview of the Satellite Power Station," 7th Intersociety Energy Conversion Engineering Conference, Conference Proceedings, September 1972, p. 712.

Morrow, W. E., Jr., Technology Review, 76, No. 2, 31 (1973).

National Petroleum Council, "U.S. Energy Outlook, A Summary Report of the National Petroleum Council," December 1972, p. 15.

National Petroleum Council, "U.S. Energy Outlook, A Report of the National Petroleum Council's Committee on U.S. Energy Outlook," December 1972.

National Petroleum Council, "U.S. Energy Outlook, An Initial Appraisal 1971-1985," July 1971.

National Petroleum Council, "Guide to National Petroleum Council Report on the United States Energy Outlook," December 1972.

National Science Foundation Legislation, Hearing before the Special Subcommittee on the National Science Foundation, May 3, 1973, U.S. Printing Office, Washington, D. C., 1973.

Nicolet, M., Arch. Met. Geophys. Biokl., B3, 209 (1951).

NSF/NASA Solar Energy Panel, "An Assessment of Solar Energy as a National Energy Resource," University of Maryland, December 1972.

Palz, W.; Besson, J.; Frémy, J.; Duy, T. Nguyen; and Vedel, J., "Analyses of the Performance and Stability of CdS Solar Cells," Conference Record of the 8th IEEE Photovoltaic Specialists Conference, Seattle, Washington, August 1970, pp. 16-23.

Paradise, M. E., U.S. Patent 2,904,613; September 15, 1959.

Pavelets, S. Yu; and Fedorus, G. A., Geliotekhnika, 7, No. 3, 3 (1971). Applied Solar Energy, 7, No. 3-4, 1 (1973).

Pulfrey, D.; and McOuat, R., Appl. Phys. Lett., 24, 167 (1974).

Queisser, H. J., Chapter 3, "Solar Cells; Outlook for Improved Efficiency," National Academy of Sciences, Washington, D. C., 1972.

Ralph, E. L., Solar Energy, 10, No. 2, 67 (1966).

Ralph, E.; and Benning, F., Conference Record of the 9th IEEE Photovoltaic Specialists Conference, Silver Spring, Mayland, 1972, pp. 370-381.

Ralph, E. L., Solar Energy, 13, 326 (1972).

Rappaport, P., RCA Rev., 20, 373 (1959).

Rappaport, P.; and Wysocki, J., "The Photovoltaic Effect," in Photoelectronic Materials and Devices, Edited by S. Larach, Van Nostrand Co., New York, 1965, pp. 239-275.

Rappaport, P.; and Wysocki, J., Acta Electronica, 5, 364 (1961).

Reddi, V.; and Sansbury, J., J. Appl. Phys., 44, 2951 (1973).

Regnier, N.; and Shaffer, M., U.S. Patent 2,919,298; December 29, 1959.

R. Riel, "Large Area Solar Cells Prepared on Silicon Sheet," Proceedings of the 17th Annual Power Sources Conference, Atlantic City, May 1963.

Rosenblatt, A. I., Electronics, April 4, 99 (1974).

Ryco Laboratories, Final Report No. AFCRL-66-134, 1965.

Savchenko, I.; and Tarnizhevskii, B., Geliotekhnika, 8, No. 4, 20 (1972). Applied Solar Energy, 8, No. 3-4, 83 (1973).

Schurr, S. H., "Energy Research Needs," Resources for the Future, Inc., Washington, D. C., October 1971, p. I-6.

Sears Fall and Winter Catalog, 1973, p. 779 (Technical Data).

Shell Oil Company, "The National Energy Outlook," March 1973.

Shearer, D., Jr.; and Farrell, R., "Battery Power," Long Range Planning Service, Stanford Research Institute, Menlo Park, Calif. 1969, Report No. 396.

Sheridan, N. R., Solar Energy, 13, 425 (1972).

Shiozawa, L.; Sullivan, G.; and Augustine, F., Conference Record of 7th Photovoltaic Specialists Conference, Pasadena, Calif., 1968, p. 39.

Shockley, W.; and Wueisser, H., J. Appl. Phys., 32, 510 (1961).

Shumann, W. A., Aviation Week and Space Technology, January 14, 56 (1974).

Solar Power Corporation, Brochure for Solar Power Module 1002, Braintree, Mass. 1974.

Spakowski, A.; and Forestieri, A., "Observations of CdS Solar Cell Stability," Conference Record of the 7th IEEE Photovoltaic Specialists Conference, Pasadena, Calif., November 1968, p. 155.

Stanley, A. G., "Degradation of CdS Thin Film Solar Cells in Different Environments," Technical Note 1970-33, Lincoln Laboratory, M.I.T., Lexington, Mass., 1970.

Stanley, A. G., "Present Status of Cadmium Sulfide Thin Film Solar Cells," Technical Note 1967-52, Lincoln Laboratory, M.I.T., Lexington, Mass., 1967.

Sterkin, C. K., "Terrestrial Applications of Solar Cells," Literature Search No. 895, Jet Propulsion Laboratory, 1968.

Tabor, H., Solar Energy, 6, No. 3, 89 (1962).

Tong, C.; and Albrecht, A., to be published, J. Chem. Phys., 1975.

Tansil, J., "Residential Consumption of Electricity 1950-1970," Oak Ridge National Laboratory, 1973, ORNL-NSF-EP-51.

Tauc, J., "Photo and Thermoelectric Effects in Semiconductors," Pergamon Press, New York, 1972, p. 18.

Technology, Chem. Eng. News, July 29, 16 (1974).

Technology Trends, Optical Spectra, September 30 (1974).

Theobald, P.; Schweinfurth, S.; and Duncan, D., "Energy Resources of the United States," Geological Survey Circular 650, Washington, 1972.

U.S. Department of Commerce, Survey of Current Business, 53, No. 12, December 1973.

U.S. Department of the Interior, "United States Energy Through the Year 2000," December 1972.

VanAerschot, A.; Capart, J.; David, K.; Fabbricotti, M.; Heffels, K.; Loferski, J.; and Reinhartz, K., "The Photovoltaic Effect in the Cu-Cd-S System," Conference Record of the 7th Photovoltaic Specialists Conference, Pasadena, Calif., November 1968, p. 22.

White, D. C., Energy Laboratory, "Final Report Submitted to the National Science Foundation—Dynamics of Energy Systems," M.I.T., Cambridge, Mass., 1973.

White, D. C., Technology Review, 76, No. 2, 11 (1973).

Williams, K. R., "An Introduction to Fuel Cells," Elsevier Publishing Company, New York, 1966, p. 314.

Wolf, M. J., "The Fundamentals of Improved Silicon Solar-Cell Performance," Chapter 4 of "Solar Cells; Outlook for Improved Efficiency," National Academy of Science, Washington, D.C., 1972.

Wolf, M., "Limitations and Possibilities for Improvements of Photovoltaic Solar Energy Converters," Proc. IRE, 48, 1246 (1960).

Wolf, M., "Historical Development of Solar Cells," 25th Power Sources Conference, Atlantic City, May 1972.

Wolf, M.; and Rauschenbach, H., "Advanced Energy Conversion,: Pergamon Press, London, 1963, pp. 455-479.

Woodall, J.; and Hovel, H., 141st Meeting of the Electrochemical Society, Houston, Texas, May 1972. Appl. Phys. Lett., 21, 379 (1972).

Wysocki, J., Solar Energy, 6, 104 (1962).

Wysocki, J.; and Rappaport, P., J. Appl. Phys., 31, 571 (1960).

Zaitseva, A.; and Polisan, A., Geliotekhnika, 8, No. 3, 28 (1972). Applied Solar Energy, 8, No. 3-4, 20 (1973).

Zarem, A.; and Erway, D., "Introduction to the Utilitization of Solar Energy," McGraw-Hill Book Co., New York, 1963, pp. 211-238.

Solow, R. [illegible] Development of [illegible] [illegible] Contributions [illegible] [illegible] [illegible]).

Smith, M. [illegible] and Kane [illegible] Advanced Energy [illegible] [illegible] 1965, [illegible]

Woodall [illegible] [illegible] of [illegible] [illegible] [illegible] [illegible] [illegible]

[illegible]

[illegible] 32 [illegible] 67 (1963).

[illegible] and [illegible] Analytic [illegible] of [illegible] [illegible] Social Theory [illegible] [illegible] (1967).

Zellner, A., and [illegible] Introduction to the [illegible] [illegible] Book Co., New York [illegible]